AMERICAN BRONZE SCULPTURE

THE NEWARK MUSEUM

Cover

Anna Vaughn Hyatt Huntington
Riders to the Sea, ca. 1912 (detail)
(Catalogue no. 38)
Photograph by William G. Wagner

AMERICAN BRONZE SCULPTURE

1850 TO THE PRESENT

GARY A. REYNOLDS

The Newark Museum
Newark, New Jersey
1984

This catalogue was published in conjunction with the exhibition, *American Bronze Sculpture: 1850 to the Present,* held at The Newark Museum from October 18, 1984 to February 3, 1985. The Newark Museum is funded by the City of Newark and the State of New Jersey. Additional funding is received from Essex County.

Published 1984
The Newark Museum
49 Washington Street
P.O. Box 540
Newark, New Jersey 07101

Library of Congress Cataloguing in Publication Data

Reynolds, Gary A.
American bronze sculpture, 1850 to the present.

Catalogue of an exhibition held at the Newark Museum, Oct., 1984-Feb. 3, 1985.
Bibliography: p.
Includes index.
1. Bronzes, American—Exhibitions. 2. Bronzes—19th century—United States—Exhibitions. 3. Bronzes—20th century—United States—Exhibitions. I. Newark Museum.
II. Title.
NB1230.R49 1984 730'.973'074014932 84-22616
ISBN 0-932828-20-5

Since its founding, The Newark Museum has collected and exhibited American bronze sculpture. In 1913 the Museum mounted *An Exhibition of Bronzes by American Sculptors* in its original rooms on the fourth floor of the Newark Free Public Library. Many of the 131 works in that show, by then-contemporary artists, are now represented in our permanent collection. A significant portion of our holdings was last on view as part of the important *Survey of American Sculpture,* organized by William H. Gerdts at the Museum in 1962.

The present exhibition and this accompanying catalogue mark an important step in The Newark Museum's tradition of research and exhibition. The pieces collected by the Museum in the past seventy-five years are seen in the context of an exhibition devoted to the American artist's fascination with the bronze medium. The final show of our seventy-fifth anniversary celebration, *American Bronze Sculpture* also heralds the planned renovation and expansion of the Museum's galleries, which will allow sculptures such as these to be on permanent view.

Curator of Painting and Sculpture Gary A. Reynolds is to be credited for his imagination and scholarship in assembling this fine catalogue and exhibition. I would also like to thank the many lenders, both public and private, whose generosity has allowed us to extend and enrich the scope of this project.

Samuel C. Miller
Director
The Newark Museum

ACKNOWLEDGMENTS

One of the most rewarding experiences of organizing any exhibition is the help received from respected colleagues. In this regard, I wish to thank the following for their efforts on behalf of this project: Douglas Baxter, Paul Cooper Gallery; Janice Conner; Linda S. Ferber, Susan Edwards and Barbara Head Millstein, The Brooklyn Museum; Ann Freedman and Bonnie Arffa, M. Knoedler & Co.; Julia Hotton, Schomburg Center for Research in Black Culture; Barbara Krulick, National Academy of Design; Susan Menconi, Hirschl & Adler Galleries; Robert Pincus-Witten; Joel Rosenkranz; Joel Shapiro, and Marjorie Strider. Very special thanks go to Herk Van Tongeren for his enthusiasm and his help in understanding the intricacies of bronze casting.

At The Newark Museum, Samuel C. Miller, Director, supported this exhibition at every turn and offered valuable advice. Mary Sue Sweeney, Director of Programs and Publications, expertly saw the catalogue through every phase of its production. Audrey Koenig, Registrar, and her assistant Teresa Al-Aref, coordinated all of the loans. David Palmer, Supervisor of the Exhibits Department, and his assistant Danny Schnur, designed a beautiful and thoughtful installation. Without the cooperation of these, and numerous other members of the staff, this exhibition truly would not have been possible.

Gary A. Reynolds
Curator of Painting and Sculpture

Edna S. Beron
The Brooklyn Museum, New York
Conner-Rosenkranz, New York
Paul Cooper Gallery, New York
Cordier & Ekstrom, Inc., New York
Terry Dintenfass Gallery, New York
The Equitable Life Assurance Society, New York
Grey Art Gallery and Study Center,
New York University Art Collection, New York
Blum Helman Gallery, New York
Hirschl & Adler Modern, New York
Malvina Hoffman Estate,
Courtesy Berry-Hill Galleries, New York
Jordan-Volpe Gallery, New York
M. Knoedler & Co., Inc., New York
Marlborough Galleries, Inc., New York
Robert Mehlman
Montclair Art Museum, New Jersey
Museum of Modern Art, New York
Mr. and Mrs. E. Jan Nadelman
National Academy of Design, New York
The New York Public Library,
The Schomburg Center for Research in Black Culture
Robert Pincus-Witten and Leon Hecht
Bernice Steinbaum Gallery, New York
Berta Walker
Whitney Museum of American Art, New York
Private Collections

From the nineteenth century to the present, bronze sculpture has had a rich tradition that underscores the interplay between artistic styles and materials. This exhibition traces the use of bronze, that special alloy of tin and copper, by American sculptors. Since it deals with only one of the many materials available to these artists, this is not intended to be a comprehensive history of American sculpture, but rather an investigation of the use of a single medium through an often complicated history.

Until the middle of the nineteenth century, most American sculpture was made of marble, wood, plaster or wax, primarily for the simple reason that American foundries did not have the technological capabilities to cast bronze sculpture. The earliest castings in this country were done during the 1840s by Robert Ball Hughes (1806-1868) and Clark Mills (1815-1883). They were responding to the increasing demand for public monuments and the realization that bronze was a more durable material than either wood or stone and better suited to survive the often harsh American climate. Both artists eventually had to construct their own foundries and spend years experimenting with casting methods. Hughes' *Nathaniel Bowditch* (1841-1847, Mount Auburn Cemetery, Boston) was so flawed that it soon had to be replaced by a cast made in Paris. With the assistance of the Prussian-born founder Charles Ludwig Richter, Mills was able to successfully complete his equestrian *Andrew Jackson* (1847-1853, Lafayette Square, Washington, D.C.), but only by casting the legs of the rearing horse in solid bronze in order to support the weight of the statue. Although neither can be considered a complete success, these two early sculptures received considerable public attention and helped foster a growing demand for outdoor bronze sculpture.

At this propitious moment the talented young sculptor Henry Kirke Brown returned from working in Italy. The artist's strong sense of nationalism led him to propound a home-grown American sculpture that would not rely on Europe. It was in this spirit that Brown constructed a foundry in 1848 in his Brooklyn studio. Although this met the needs for casting the small sculptures that he modeled during the first years after his return to New York, it could not handle Brown's first important public commission, a life-size equestrian statue of George Washington for the City of New York. The artist was able to convince James Tyler Ames, director of the Ames Manufacturing Company in Chicopee, Massachusetts, to take on the work. Although the Ames foundry had cast for the artist as early as 1851, this was by far their most complex job. Brown and his assistant, John Quincy Adams Ward, spent many months at the foundry during 1854 and 1855 supervising the workmen. Before its association with Brown, the Ames foundry had no experience in the fine arts and was chosen for its reputation for casting large cannons and military weapons.

The great acclaim that greeted Brown's statue of Washington when it was unveiled in Union Square on July 4, 1856, firmly established the reputations of both artist and foundry. For almost three decades Brown was the most influential sculptor working in this country, a mantle that passed at his death in 1886 to his former student, J.Q.A. Ward. The Ames Manufacturing Company occupied a similar position in the casting of bronze sculpture, although its most important work was done by 1875. Perhaps most significantly, the success of Brown's *George Washington* encouraged other commercial founders to turn to casting sculpture. Although not all could meet the standards of the Ames company, Robert Wood and Company (Philadelphia), George Fishcher and Brother (New York), Bureau Brothers (Philadelphia) and The Henry-Bonnard Bronze Company (New York) were among the many foundries that grew up in the second half of the nineteenth century to meet the new demand for bronze sculpture in America.

Until the late 1890s, the only method practiced by American bronze founders was sand casting. Simply described, the process begins with the casting of a plaster duplicate of the artist's original model, which would most often have been in clay. This plaster is then used to make a negative impression in a two-part mold of compressed sand. A core of plaster-like material is suspended by means of an armature inside the joined halves of the sand mold. A system of vents (for the escape of air and gasses) and gates (for the flow of bronze) are cut into the mold. Next, molten bronze is poured into a hole at the top of the mold, filling the space between the negative impression and the core. After the bronze hardens, the sculpture is removed from the mold (which is destroyed in the process) and the core knocked out. The extraneous pieces from the armature and vents are cut off; the work is chased, finished and patinated (colored).

Sand casting is a relatively inexpensive method of producing bronze sculpture but it does have limitations. It is extremely difficult to build a sand mold that will retain its form unless the surface of the sculpture is smooth and free from deeply undercut details. Large or intricately modeled sculptures have to be cast in several pieces and assembled by the sculptor or his assistants. Since each sand mold can only be used once, the method is most practical for unique castings or for a very small number of replicas. These considerations aside, however, the reason nineteenth-century American founders relied on sand casting was that they knew no other method.

It was not until 1897 that two Italian immigrants, Riccardo Bertelli and Giuseppe Moretti, established the first foundry in the United States to do lost-wax casting. This method fell into disuse in the early eighteenth century and its ancient techniques were forgotten until revived by European, primarily French and Italian, founders in the mid-nineteenth century. Like sand casting, the lost-wax method begins with a plaster, cast from the artist's original model. Next, the plaster model is placed in a retainer mold and liquid gelatin or glue poured around it. This creates a flexible mold that bears an exact negative imprint of the artist's model. It is into this mold that a thin coating of wax is brushed or poured. The hollow center of the wax impression is filled with a core material and a plaster base with drain lines is built. The wax model is then removed from its gelatin mold and a system of gates, vents and channels (all composed of wax) is added to the exterior. This piece is then encased in a plaster or ceramic mold and fired in a kiln until all the wax has drained from it. Molten bronze is poured into the mold and fills the spaces formerly occupied by the wax. After the bronze has hardened, the mold is knocked off, the core removed, and the extraneous runners and vents cut off. The sculpture is then ready to be chased, finished and patinated.

Most American sculptors saw a number of advantages to the lost-wax process. Unlike sand, gelatin could penetrate every crevice of the artist's model. Combined with its flexibility, this produced a mold that could duplicate the most intricately detailed and undercut surfaces. In addition, the surface of the wax model could be retouched or slightly altered before being encased in its plaster mold. Since the gelatin mold could be reused after each wax impression, for the first time it became practical to reproduce bronze sculpture in relatively large editions. A final advantage to the process was that the flexible gelatin mold enabled most sculptures, unless they were exceptionally large, to be cast in one piece. The foundry that Bertelli and Moretti established, which later became the Roman Bronze Works, revolutionized the business of bronze casting in America. By the early years of the twentieth century, the American sculptor could have his works cast at several foundries practicing the lost-wax method, including Roman Bronze Works (New York), the Gorham Manufacturing Company (Providence, Rhode Island), E. Gargani & Sons (New York), the California Art Foundry (Los Angeles) and the Modern Art Foundry (New York).

These technological advances in bronze casting coincided with major changes in the aesthetics of nineteenth-century American sculpture. From the mid-1820s through the mid-1870s, American sculpture was dominated by the

Neoclassic style. The leaders of this movement, Horatio Greenough (1800-1852), Hiram Powers (1805-1873) and Thomas Crawford (ca. 1813-1857), spent their careers in Italy, drawing inspiration from antique Greek and Roman sculpture. All upheld the primacy of the classical tradition, with its idealized proportions and clarity of form. Although a typical piece of Neoclassic sculpture, such as Powers' famous *Greek Slave* (1843), first took shape in the artist's clay model, its final apotheosis was in pure white marble, carved by the artist's assistants. This aesthetic was carried into the second half of the nineteenth century by a younger generation of Neoclassic sculptors led by Randolph Rogers (1825-1892), William Wetmore Story (1819-1895) and William Henry Rinehart (1825-1874), all of whom lived as expatriates in Italy. Even those sculptors such as Henry Kirke Brown, John Quincy Adams Ward or Thomas Ball, who chose to work at home and promoted what they felt to be a more "American" style of naturalism, still owed a strong debt to Neoclassicism.

As Paris became the mecca for young American art students in the late 1870s, new ideas challenged the primacy of Neoclassicism. In private ateliers, in the classrooms of the Ecole des Beaux-Arts and in the annual Salon exhibitions, young Americans intent on becoming sculptors were confronted with a new kind of sculpture, and most often it was cast in bronze. Earlier, the Romantics, such as Antoine-Louis Barye (1796-1875) and Pierre-Jean David D'Angers (1788-1856), had helped to create a new taste in French sculpture for naturalism and flamboyant modeling. Founders revived the art of lost-wax casting to successfully duplicate the lively surfaces and intricate details of the artist's original clay model. By the 1880s, when American students began to arrive in Paris in large numbers, the emphasis in French sculpture had shifted from carving to modeling; in effect, from marble to bronze.

The aim of most of these students was to study at the prestigious, state-run Ecole des Beaux-Arts. Although entrance to the Ecole required some proof of proficiency in drawing, reforms instituted in 1863 gave the individual instructors final decision as to who would be admitted into their classes. The atelier of Francois Jouffroy (1806-1882), and later those of his former students Jean-Alexandre-Joseph Falguière (1831-1900) and Marius-Jean-Antonin Mercié (1845-1916), were favorite choices of American sculptors entering the Ecole. However, art instruction in Paris was not limited to that institution. Artists could receive instruction, free of tuition, in anatomy, modeling and drawing at the less-prestigious Petite Ecole (Ecole Gratuite de Dessin). In addition, there were the private schools, most notably the Académie Julian and the Académie Colarossi, that hired prominent artists, such as the sculptor Henri Chapu (1833-1891), to teach and offer criticism in its atelier-based system. Young sculptors with a special interest in animal subjects could study anatomical drawing in the classes offered at the natural history museum of the Paris zoological garden, Les Jardin des Plantes, and taught successively by the two great *animaliers* Antoine-Louis Barye and Emmanuel Frémiet (1824-1910). And, finally, there were numerous artist's ateliers, or studios, that took students.

Returning from their studies in Paris in the last years of the nineteenth century, American artists brought back a new style of bronze sculpture, one that stressed expressive modeling, a heightened sense of naturalism and active, often complex, surfaces. They helped found new organizations to promote and exhibit their sculpture, chief among them the National Sculpture Society (1893) and the Architectural League of New York (1881). In addition, these artists helped disseminate the styles and techniques learned abroad through teaching in the art schools burgeoning throughout the United States, such as the Art Students League of New York (1875). The great international fairs–World's Columbian Exposition (Chicago, 1893), Pan American Exposition (Buffalo, 1901), Louisiana Purchase Exposition (Saint Louis, 1904) and Pamama-Pacific International Exposition (San Francisco, 1915)–were all conceived with large sculpture programs that provided employment. No longer dependent upon European foundries for the casting of their work, American sculptors returned to this country with far greater frequency after 1900 to establish their own studios.

Even as styles of sculpture continued to change during the early years of the twentieth century, bronze remained an important medium. The work of Auguste Rodin (1840-1917) became a pervasive influence on the young artists who continued to study in Paris during the early years of this century. The sketchy, incomplete look of Rodin's sculpture prompted Americans such as Malvina Hoffman to further investigate the fluid, expressive qualities of bronze.

At no time during the twentieth century has bronze sculpture recaptured the popularity it achieved with Americans prior to World War I. By the 1920s, it had seriously lost its appeal for progressive American artists. The desire for large public monuments, which in the nineteenth century had helped spur American interest in bronze sculpture, soon became embarrassingly old-fashioned and passé. In addition, bronze sculpture was too closely associated with the academic or Beaux-Arts style to provide inspiration for newer generations of twentieth-century artists. In their eagerness to break with the past, many early modern sculptors sought new materials less bound to traditional values. William Zorach, John B. Flannagan and Robert Laurent promoted a modernist aesthetic that emphasized the direct involvement of the artist with his materials. Carving produced a unique object, direct from the artist's hands, that was the antithesis of bronze sculpture produced in editions of exact replicas.

This is not to say that early twentieth-century sculptors entirely rejected the use of bronze. Elie Nadelman, one of the most inventive artists of this century, had many of his important works cast in bronze. However, these occupy a smaller portion of his oeuvre than the numerous sculptures executed in wood, stone, terra-cotta and ceramic. The same generalization holds true for many leading abstract sculptors active during the 1930s, including Alexander Archipenko, Gaston Lachaise, John Storrs, Isamu Noguchi and Alexander Calder. It was most often the figurative, academically-oriented artists of the period, like Paul Manship, who continued to have the majority of their work cast in bronze.

During the 1940s and 1950s, bronze sculpture appears to have made a short-lived comeback in the sculptor equivalents of the Abstract Expressionist painters. Seymour Lipton, Herbert Ferber and David Hare all exploited the expressive qualities of the medium in their sculpture. Like their immediate predecessors, these artists usually rejected the use of bronze for replication and emphasized the directness of their work in unique castings. David Smith, perhaps the best-known sculptor of the period, cast some of his early works in bronze but then moved on to the sculptures of welded steel upon which his reputation and influence are based. The names of sculptors such as Joseph Cornell and Louise Nevelson are entirely absent from the history of bronze sculpture.

The next two decades of American sculpture saw an even further diminishing of the number of artists working in bronze. The industrial processes of welding and fabrication took on added significance for sculptors of the 1960s and 1970s. Not only industrial processes but also industrial materials helped form the aesthetic of Minimal sculpture. Bronze seemed the least appropriate material for a generation of artists who strove to eliminate the individual's "touch" from their work. The aluminum and steel boxes of Donald Judd, the neon constructions of Dan Flavin, the flat copper and zinc floor plates of Carl Andre and the large steel constructions of Tony Smith or Mark di Suvero shape our view of what sculpture has meant during the last two decades. This view has rarely included bronze sculpture.

It is only during the last few years that bronze has again become of major interest to American sculptors. No doubt this has much to do with the interest in new or unusual materials and the resurgence of self-expression evident in so much contemporary art. Although the basic techniques of casting have changed very little since the nineteenth century, a number of new materials to patinate, or color, bronze sculpture have been developed. Strictly speaking, patination refers to any natural or artificially-induced alteration of the surface of the bronze, which usually emerges from casting a bright golden color. Traditionally, artists have applied chemical solutions under heat to achieve a range of green and brown hues. Today, in addition to simply painting the surface, artists can choose from chemical processes that produce an astounding array of colors. No longer limited to traditional greens and browns, bronze sculpture can now be made to look like almost any other material.

Whether cast in large editions or as unique pieces, traditionally patinated or colored to resemble other materials, cast or welded, bronze sculpture is very much a part of contemporary art. Although few artists work exclusively in the material, bronze sculpture has gained a new popularity that is reminiscent of its position at the turn of the century. In this respect, the history of bronze sculpture in America has come full circle.

The catalogue is arranged chronologically by birth date of the sculptor; an alphabetical index of the artists appears on page 71. All dimensions are given in inches. Height precedes width (length) and depth. Dates refer to the year (or years) in which the sculpture was modeled. When the date of casting is known it has been included. In the location of the marks, the proper right and proper left of the sculpture are abbreviated as "pr" and "pl".

HENRY KIRKE BROWN
Born 1814, Lyden, Mass.
Died 1886, Newburgh, N.Y.

Brown's first training as an artist was with the Boston portrait painter Chester Harding. He began modeling clay sculptures while in Cincinnati about 1837, probably with the encouragement of the local sculptor Schobail Clevenger. After a year painting and modeling portrait busts in Albany, New York, Brown returned to Boston in preparation for a trip to Italy. The artist remained in Florence for approximately a year and then moved to Rome. While working in Rome, the intuitive realism of Brown's early work was replaced by a more idealized neoclassicism. Brown decided to return to the United States and, in 1846, set up his studio in New York. By 1848, the studio was moved to Brooklyn where, responding to the inability of commercial founders to make suitable castings, Brown built his own small bronze foundry. Two small bronzes that he cast in editions of twenty for distribution by the American Art-Union in 1849 and 1850 are thought to be the first multiple castings done in the United States. He spent most of 1853 to 1856 working on the important commission for the equestrian statue of *George Washington* (Union Square, New York) and the supervision of its casting at the Ames Foundry in Chicopee, Massachusetts. In 1857, Brown moved his studio to Newburgh, New York, and spent the second half of his prolific career executing numerous commissions for public sculpture.

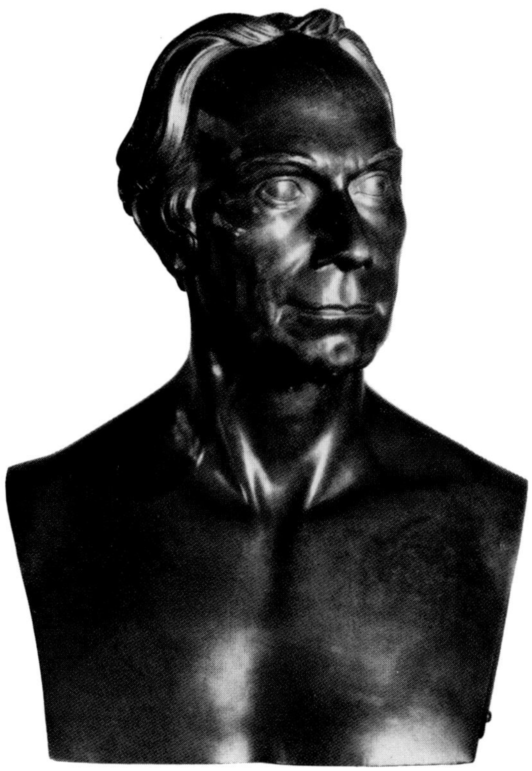

1.
Henry Clay
1852

Cast bronze, golden brown patina, H. 16¼″ (including integral base)

Signed and dated: *H.K. Brown Sculptor/June 1852* (on pr shoulder)

No foundry mark (probably cast in artist's studio)

Collection of The Newark Museum. Bequest of Dr. J. Ackerman Coles 1926 (26.1044)

THOMAS BALL
Born 1819, Charlestown, Mass.
Died 1911, Montclair, N.J.

Thomas Ball's early career was as a painter. With no formal training, he began making small sculptures in 1850 and 1851. Encouraged by the success of his first sculptures, Ball and his wife sailed for Europe in 1854, eventually settling in Florence. There, they became part of the American expatriate community that included Hiram Powers, Joel T. Hart and Thomas B. Read. Ball returned to Boston in 1857 and remained for almost eight years. His most famous work from this period is the bronze equestrian statue of *George Washington* in the Boston Public Garden. After his return to Florence in 1864, Ball executed numerous portrait busts in marble and, in the mid-1870s, several heroic-size bronze statues (*Charles Sumner*, 1878, Boston Public Garden; *Daniel Webster*, 1876, Central Park, New York). From 1897 until his death at the age of ninety-two, Ball lived with his daughter and her husband, the sculptor William Couper.

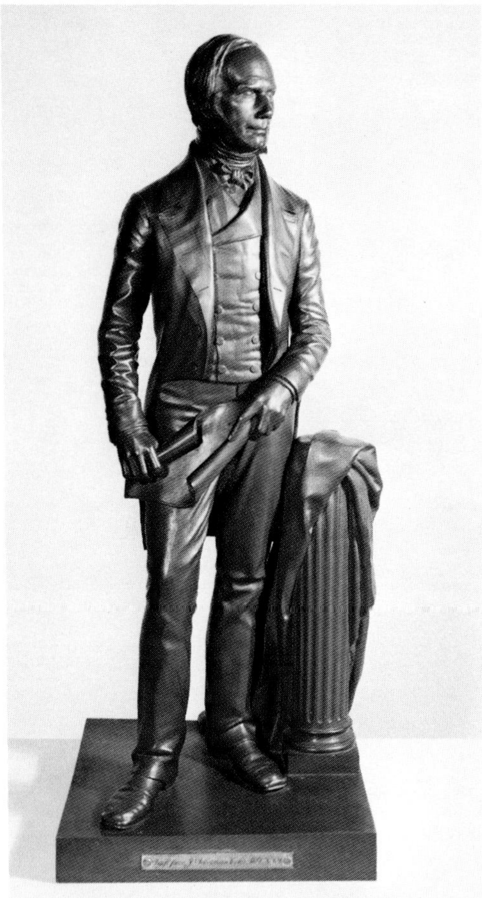

2.
Henry Clay
1858

Cast bronze, green-brown patina, 30½" x 11½" x 11"

Signed and dated: *T. Ball Sculp. Boston 1858* (on drapery on back)

Inscribed: *Patent assigned to G.W. Nichols*

No foundry mark (probably cast at Ames Foundry, Chicopee, Mass.)

Collection of The Newark Museum. Gift of Dr. J. Ackerman Coles 1919
(19.718)

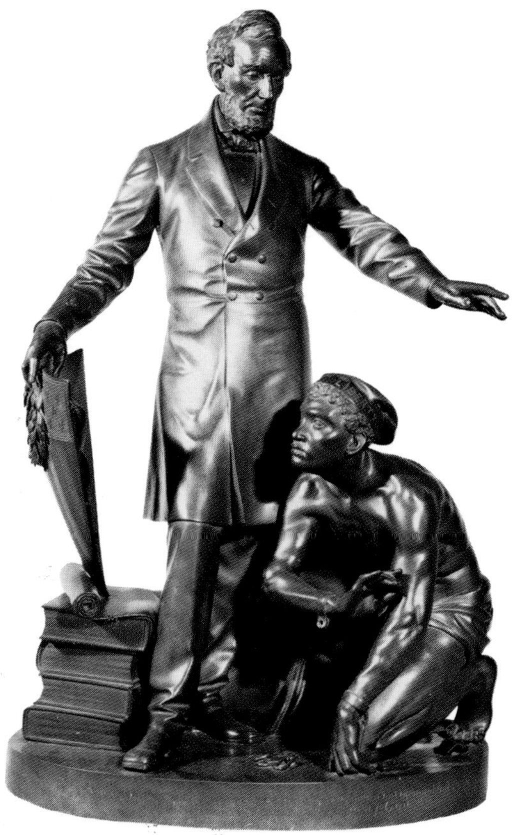

3.
Emancipation Group
1865; cast 1873

Cast bronze, green-brown patina, H. 34½"

Signed and dated: *T. Ball Sculpt/1865* (on base at right)

Foundry mark: *Fausto Favi Fuse/ Firenze 1873* (on base at left)

Inscribed: *And upon this act–I invoke the considerate judgment of mankind/and the gracious favor of Almight God* (on front of base)

Lent by the Montclair Art Museum. Gift of Mrs. William Couper

This is a reduced version of the life-size monument erected in Washington, D.C. (1875) and Boston (1877).

JOHN QUINCY ADAMS WARD
Born 1830, Urbana, Ohio
Died 1910, New York City

Unlike most other sculptors of his generation, Ward never traveled to Europe for study. His training was in the studio of Henry Kirke Brown, where he worked from 1849 to 1857 as both student and apprentice. Much of his practical experience in bronze casting was learned while assisting Brown with his equestrian statue of George Washington. Ward worked briefly in Washington, D.C. and Columbus, Ohio before establishing his own studio in New York in 1861. Some of his earliest work was as a designer of decorative objects for the Ames Foundry, with whom he had worked on the casting of Brown's *Washington*. Although Ward was often praised for the straightforward naturalism of his sculpture, he was always open to other influences. Generally modeled from life, his early figures nevertheless incorporate poses and small details from antique sculpture. Ward became one of the most successful of American sculptors, serving as president of both the National Academy of Design and the National Sculpture Society. Many of his bronze statues adorn the parks and public spaces of New York City (*The Indian Hunter*, 1869, Central Park Mall; *George Washington*, 1883, Wall and Broad Streets; *Horace Greeley*, 1890, City Hall Park).

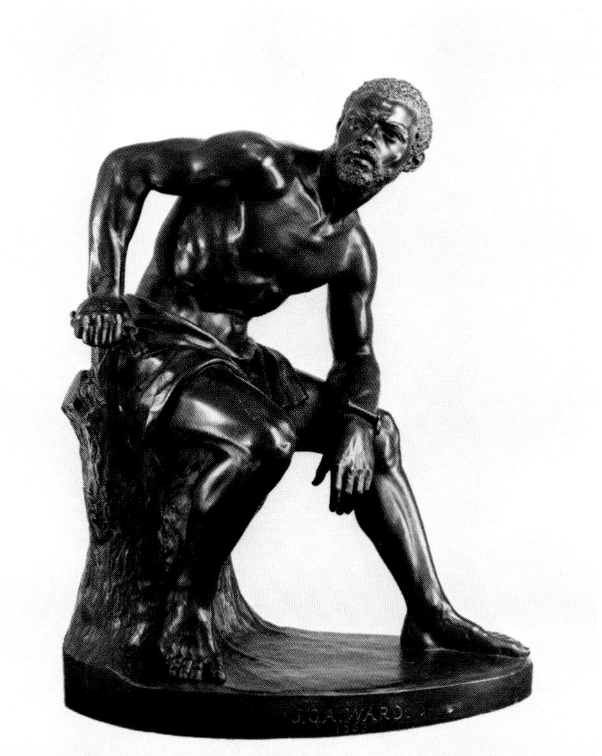

4.
Freedman
1863; cast 1899

Cast bronze, dark green-brown patina, 20″ x 14¾″ x 7″

Signed and dated: *J.Q.A. WARD. sc/1863* (on front of base)

Foundry mark: *The Henry-Bonnard Bronze Co/Founders. N-Y. 1899.* (on base at back)

Lent by the National Academy of Design, New York

HARRIET HOSMER
Born 1830, Watertown, Mass.
Died 1908, Watertown, Mass.

Her independent nature and mannish attire made Hosmer one of the most colorful of what Henry James called "the white marmorean flock," the group of American women sculptors working in Italy during the mid-nineteenth century. Before settling in Rome in 1852, Hosmer had taken lessons from the Boston sculptor Paul Stevenson, studied anatomy at Washington University in St. Louis, and learned to carve her own marble sculpture. She was able to convince the celebrated English sculptor, John Gibson, then working in Rome, to accept her as a student. Working with Gibson and copying antique sculpture, Hosmer became thoroughly imbued with the Neoclassical aesthetic. The artist's coterie of friends included many of the leading American and European writers, actors and artists. Her sculpture of the clasped hands of Elizabeth and Robert Browning is her only known work in bronze and a touching tribute to her friendship with the couple. Typically, she cast the several known replicas herself.

5.
Hands of Robert and Elizabeth Barrett Browning
1853

Cast bronze, brown patina, 3¼″ x 8¼″ x 4½″

Signed and dated: *Hands of Robert/and/Elizabeth Barrett Browning/cast by Hosmer/Harriet/Rome 1853* (on wrist of male hand)

Inscribed: *Copyright* (on wrist of female hand)

Collection of The Newark Museum. Gift of Mrs. J.G. Phelps Stokes 1976 (76.8)

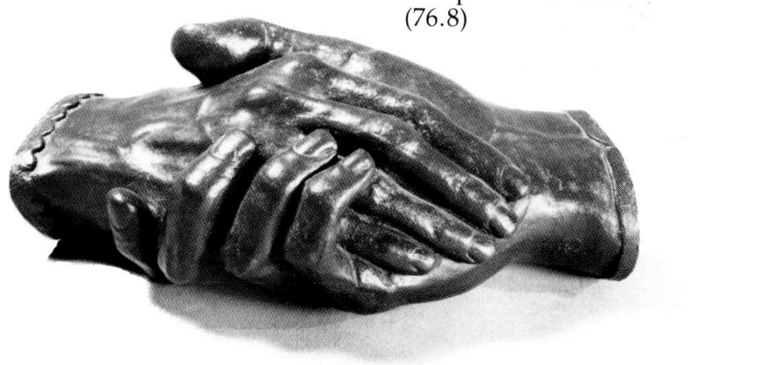

OLIN LEVI WARNER
Born 1844, Suffield, Conn.
Died 1896, New York City

Warner is representative of the younger generation of American sculptors who began to seek instruction in Paris, rather than in Rome, during the late 1860s. After deciding to become a sculptor at the age of nineteen and working at various jobs to earn the necessary money, Warner arrived in Paris in 1869. He entered the Ecole des Beaux-Arts and studied in the atelier of Francois Jouffroy. A talented and industrious student, Warner worked for a short time as a studio assistant for the successful French sculptor Jean-Baptiste Carpeaux. His friends in Paris included Augustus Saint-Gaudens, Jean-Alexandre-Joseph Falguière and Marius-Jean-Antonin Mercié, all fellow students of Jouffroy. Warner mastered the heightened naturalism and sensuous surfaces being taught by his French instructors but upon his return to the United States in 1872 had difficulty finding patrons for his sculpture. It was not until the late 1870s that his sensitively modeled portrait busts and medallions began to win him fame and commissions. The artist died suddenly from a bicycle riding accident in Central Park while engaged on his most ambitious project, a pair of bronze doors for the Library of Congress.

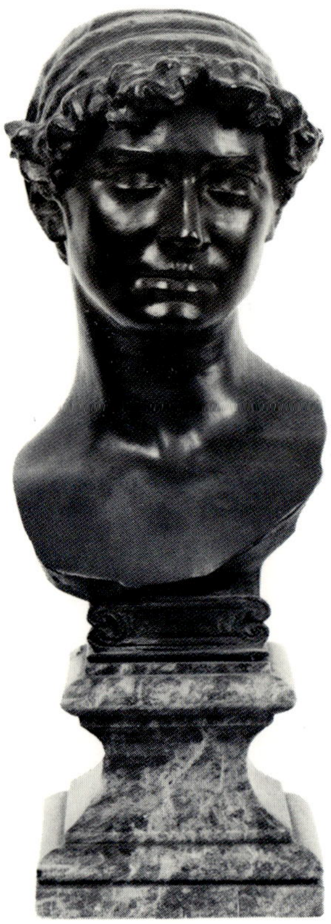

6.
Portrait of a Woman (*Maud Morgan*)
1880

Cast bronze, dark brown patina,
H. 23½″

No markings

Lent by The Brooklyn Museum.
On loan from Mrs. Olin Warner

Maud Morgan (1861-1941) was an American musician, harpist, teacher and author.

AUGUSTUS SAINT-GAUDENS
Born 1848, Dublin, Ireland
Died 1907, Cornish, N.H.

Saint-Gaudens was one of the first artists to popularize the French Beaux-Arts style of sculpture in the United States. The artist grew up in New York City and at an early age was apprenticed to a cameo cutter. At the age of nineteen he went to Paris, where he studied drawing at the Petite Ecole in order to be admitted to the Ecole des Beaux-Arts, where he worked in Jouffroy's atelier. For five years (1870-75), Saint-Gaudens worked primarily in Rome absorbing some of the idealization of antique sculpture and developing an interest in the fluidly modeled surfaces of fifteenth-century Florentine bronzes. Shortly after returning to the United States in 1875, he received the important commission for the Farragut Monument. The figure and accompanying bas reliefs were modelled in Rome and Paris and unveiled in Madison Square, New York City, in 1881. For the rest of his career the artist worked in his New York studio and at his farm in Cornish, New Hampshire. Saint-Gaudens became one of the most influential artists of his time and his public sculptures include *Abraham Lincoln* (1887, Lincoln Park, Chicago), the *Shaw Memorial* (1896, Boston), the figure of *Diana* (1892) that stood atop the original Madison Square Garden and the equestrian *Sherman Monument* (1903, New York). During his career, the artist worked on numerous projects with the architects Stanford White and Charles McKim.

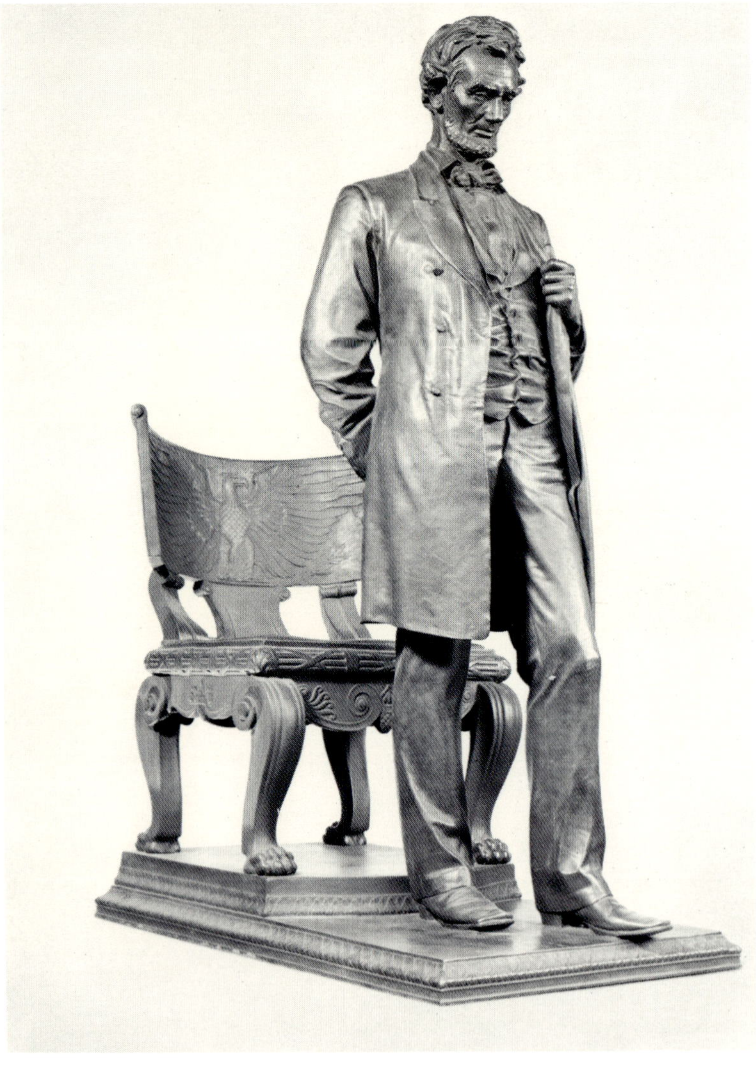

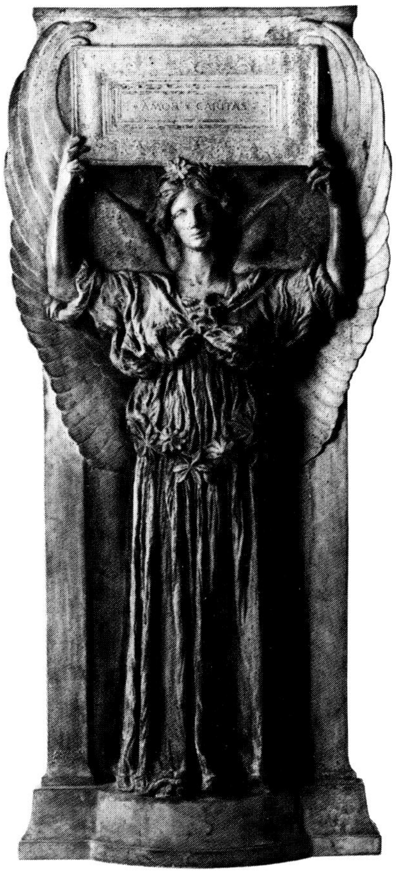

8.
Amor Caritas
1898

Cast bronze, green patina, 39⅞" x 17½" x 4"

Signed and dated: *Augustus/ Saint-Gaudens/MDCCCXCVIII* (on front at lower left)

Inscribed: *Amor • Caritas* (on front in plaque)

Lent by The Brooklyn Museum. Gift of Mr. and Mrs. Harris Klein

9.
Victory
1904

Cast bronze, golden brown patina, H. 8"; green marble base, 4½" x 4" x 4"

Signed and dated: *A.S.T. Gaudens.Aspet.MCMIV* (under pl neck)

Inscribed: *Victory–Peace* (in Greek on plaque on front)

No foundry mark

Collection of The Newark Museum. Purchase 1927 Dr. J. Ackerman Coles Bequest Fund (27.287)

This is a cast of a second study (not used) for the head of "Victory" for the *Sherman Memorial*, unveiled in 1903 in Grand Army Plaza, New York City.

7.
Abraham Lincoln
1887

Cast bronze, golden brown patina, 40½" x 15¾" x 27½"

Signed and dated: *Augustus. Saint-Gaudens-Sculptor. M.D.C.C.C.LXXXVII* (on base at right)

Inscribed: *Copyright.1912.by.A.H. Saint.Gaudens* (on back of base)

Foundry mark: *Gorham Co. Founders/*[hallmark]*/OAWM* (stamped on base near back left leg of chair)

Collection of The Newark Museum. Gift of Franklin Murphy, Jr., in memory of Franklin Murphy 1920 (20.135)

This is one of several posthumous reductions of the monumental statue in Lincoln Park, Chicago.

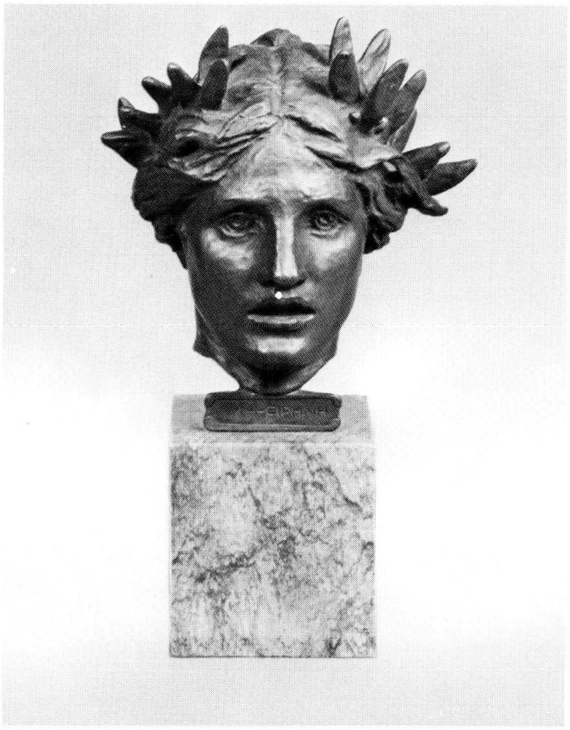

DANIEL CHESTER FRENCH
Born 1850, Exeter, N.H.
Died 1931, Stockbridge, Mass.

Ultimately achieving even greater fame than Warner or Saint-Gaudens, French helped revitalize American sculpture and the use of bronze at the turn of the century. Growing up in Cambridge and Concord, Massachusetts, the artist briefly attended Massachusetts Institute of Technology, but in 1870 his interest in sculpture led him to New York and a month's apprenticeship in J.Q.A. Ward's studio. With only this meager training, French was awarded the commission by the town of Concord for a monument commemorating the American centennial. French began modeling his *Minute Man* in 1873, taking additional study in anatomy with William Rimmer, and sent it for casting at the Ames Foundry two years later. The next eighteen months were spent in Europe, working in Thomas Ball's studio in Florence. After returning to the United States, French began to regret his, by then, old-fashioned dependence on the Neoclassical tradition. In 1886, he left for study in Paris, where he enrolled at the Académie Julian. During the year he spent in Paris, French's work became more assured in its modeling, complex in composition and lively in surface treatment. The artist created many bronze public sculptures throughout his long career, including the *Thomas Gallaudet Memorial* (1888, Gallaudet College), *John Harvard* (1884, Harvard University), *Alma Mater* (1903, Columbia University) and the standing *Abraham Lincoln* (1912, Lincoln, Nebraska). Undoubtedly French's most famous sculpture is the marble seated Lincoln in the Lincoln Memorial, Washington, D.C. Like Saint-Gaudens, French worked with the leading architects of his day, including Cass Gilbert (for whose U.S. Custom House in New York he created entrance sculptures of the four Continents) and the firm of McKim, Mead and White. In 1897 the artist bought property in the Berkshire Mountains near Glendale, Massachusetts, where he maintained his summer home and studio. Known as Chesterwood, this property is now open to the public, administered by the National Trust for Historic Preservation.

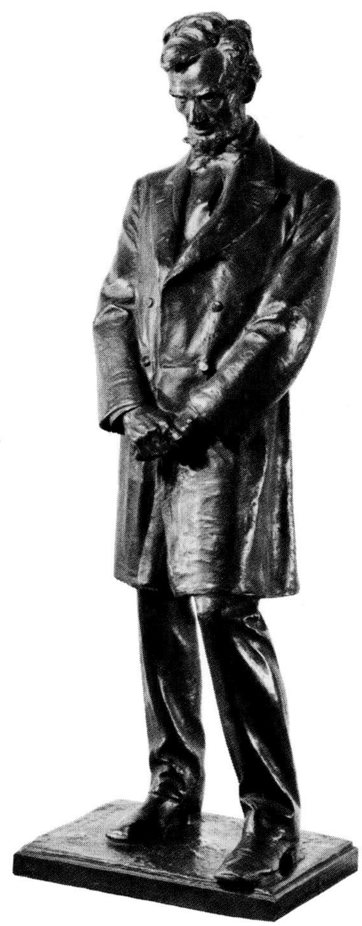

11.
Abraham Lincoln
1912

Cast bronze, brown patina, H. 37½″

Signed and dated: *Daniel C French Sc/1912* (on base at left)

Inscribed: *Original Model for the Statue Erected in Lincoln, Nebraska* (on base at right)

Foundry mark: *Cast by Roman Bronze Works. N.Y.-* (on back of base)

Collection of The Newark Museum. Purchase 1913 (13.281)

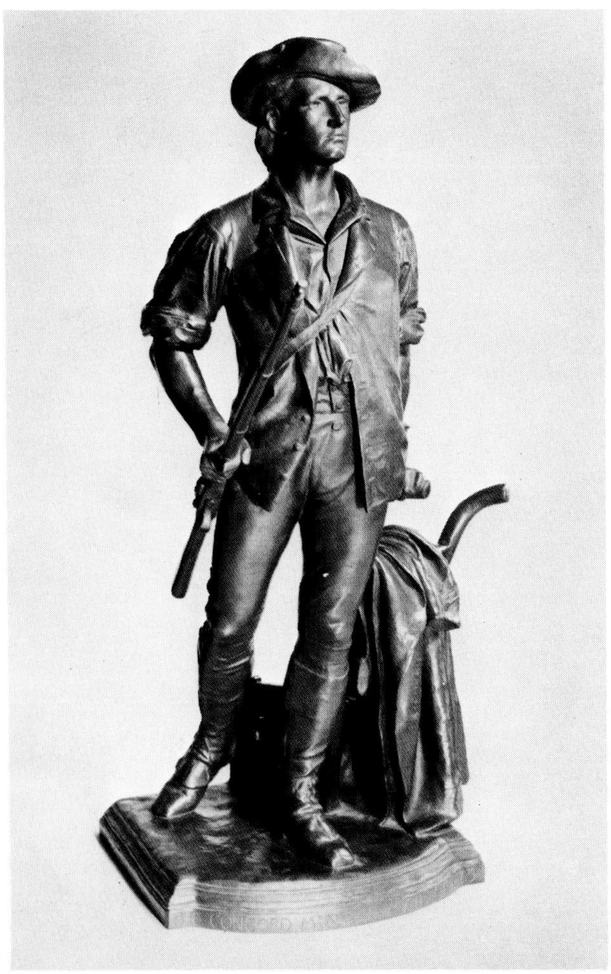

10.
The Concord Minute Man of 1775
1889-90

Cast bronze, brown patina, 32¼″ x 16″ x 20½″

Signed: *D.C. French* (on base at left)

Inscribed: *The Concord Minute Man of 1775* (on front of base)

Foundry mark: *Gorham Co. Founders*/[hallmark]/*OARS″*

Collection of The Newark Museum. Bequest of Miss Annie W. Colby 1960 (60.588)

In 1889-90, French reworked his statue of the *Minute Man* in a reduced size for installation on the gunboat *Concord*. The original casting of this version was by M.H. Mosman, Chicopee, Massachusetts. Beginning in 1913, Gorham Founders made numerous replicas of this piece.

12.
The Spirit of Life
1914

Cast bronze, gilded,
48¾" x 32" x 36½"

Signed and dated: *D.C. French/ May • 1914* (on rock)

Foundry mark: *Roman Bronze Works N.-Y.-* (on base at back)

Collection of The Newark Museum. Gift of Mr. and Mrs. Felix Fuld 1927
(27.1151)

This is a cast of the working model for the Spencer Trask Memorial Fountain in Saratoga Springs, New York. It is one of six replicas cast from 1923 to 1931.

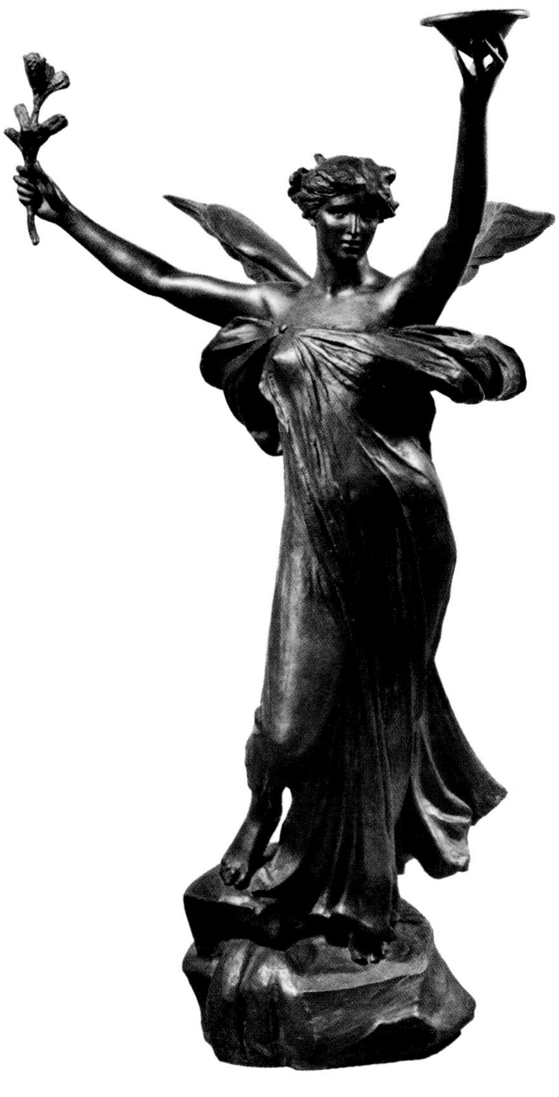

JOHN TALBOTT DONOGHUE
Born 1853, Chicago, Ill.
Died 1903, Hamden, Conn.

By the 1880s, Paris had become a mecca for young American sculptors. Donoghue worked there from 1875 until 1881, after winning a scholarship from the Chicago Academy of Design. He studied at the Ecole des Beaux-Arts under Jouffroy and later, after a short return to the United States, with Falguière. From 1884 to 1887 the artist worked in Rome, where he often returned in later years. Despite the success of works such as *The Young Sophocles*, which Lorado Taft classed in 1903 as "among the most perfect examples of ideal sculpture yet produced by an American," Donoghue was largely unsuccessful in receiving important public commissions. In the 1890s, he worked in New York as an architectural sculptor and examples of his work are to be seen on the facades of the Appellate Court Building in New York and the Library of Congress in Washington. The rejection of his design for the McKinley Monument in Philadelphia was a great blow to the artist's self-esteem and eventually led to his suicide.

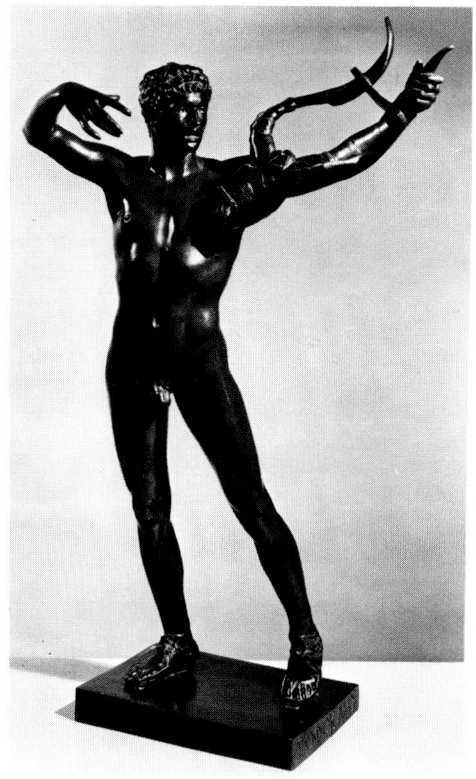

13.
The Young Sophocles
(The Young Sophocles Leading the Chorus of Victory after the Battle of Salamis)
1885

Cast bronze, medium brown patina, H. 44½"

Signed: *J. Donoghue.SC.* (on base)

Inscribed: *Sophocles Salamis* (in Greek, on base)

No foundry mark

Private Collection

A large version of this piece (H. 92½") was exhibited at the Paris Salon of 1886, where it won an honorable mention. In 1893 it was awarded a first prize at the World's Columbian Exposition.

HERBERT ADAMS
Born 1858, West Concord, Vt.
Died 1945, New York City

After studying at the Massachusetts Normal School of Art in Boston, Adams traveled to Paris in 1890 for further study. For almost five years he worked in the atelier of Marius-Jean-Antonin Mercié, one of the most successful academic sculptors of his day. From Mercié, Adams acquired a taste for Florentine Renaissance sculpture, whose softly undulating forms are often found in his own work. Adams also adopted the popular French technique of painting or tinting marble and adding details of wood, metal and other materials. This highly decorative combination of materials characterized a series of delicate female portrait busts for which he received much praise. After his return to the United States, Adams taught at the newly-established Pratt Institute in Brooklyn, remaining on the faculty until 1898. Among his most important public commissions were monumental bronze statues of *William Cullen Bryant* (1911, Bryant Park, New York City) and *William Ellery Channing* (Boston Public Garden). Adams was closely associated with many conservative art organizations and served as president of both the National Academy of Design and the National Sculpture Society.

JOHN MASSEY RHIND
Born 1860, Edinburgh, Scotland
Died 1936, London, England

Rhind was a trained sculptor when he immigrated to the United States in 1889. A third-generation member of a family of Edinburgh sculptors, he had studied in London and Paris, primarily with Jules Dalou. Rhind's first employment was as a designer of decorative sculpture for buildings. Along with Karl Bitter and Charles Niehaus, he won the important competition for one of the three bronze doors given by the Astor family to Trinity Church, New York. This competition, for which Rhind designed panels depicting the expulsion of Adam and Eve, was often likened to that carried out during the Renaissance for the Baptistry in Florence. Rhind was an important participant in the turn-of-the-century movement called the "American Renaissance," which brought together architects, painters and sculptors for grand public and private buildings. He designed sculpture to adorn the American Security Building and the Exchange Court Building in New York, Alexander Commencement Hall at Princeton University, and numerous public monuments, including perhaps his most successful statue, *George Washington* (1912), in Newark's Washington Park. To accommodate the large demand for his work, the artist maintained studios in New York City and in Alpine, New Jersey.

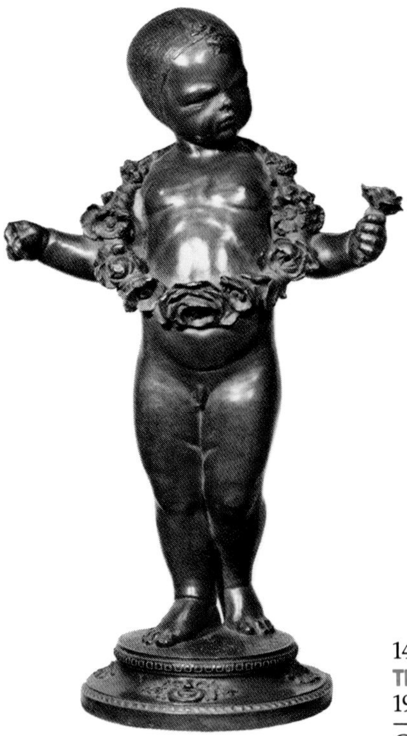

14.
The Infant Burbank
1905

Cast bronze, brown patina, H. 25½"

Signed and dated: *Herbert Adams/MCMV/©* (on rear of base beneath pl foot)

Foundry mark: *Roman Bronze Works N-Y-* (on back of base)

Collection of The Newark Museum. Purchase 1916 (16.472)

A replica of this statue, a tribute to the American horticulturist Luther Burbank (1849-1926), was kept by the artist as a fountain figure in his own garden.

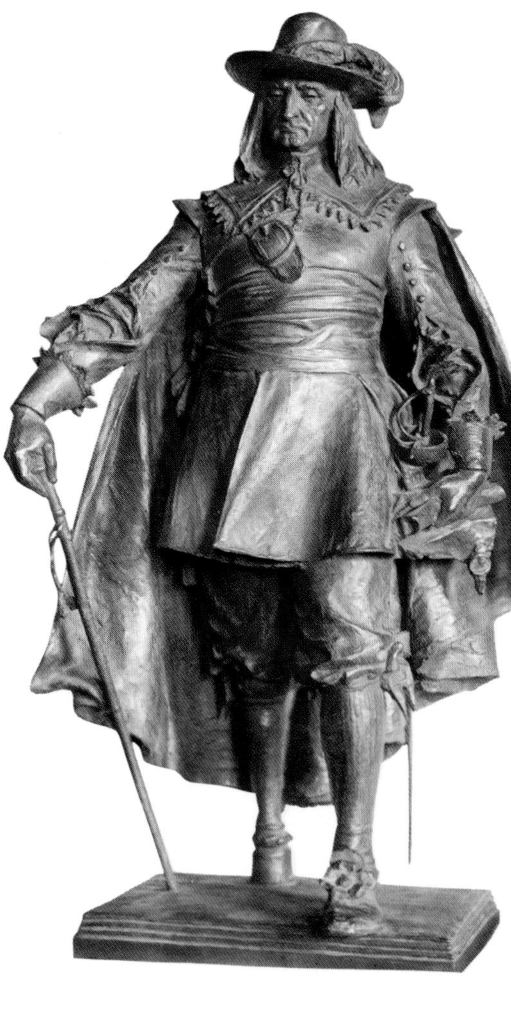

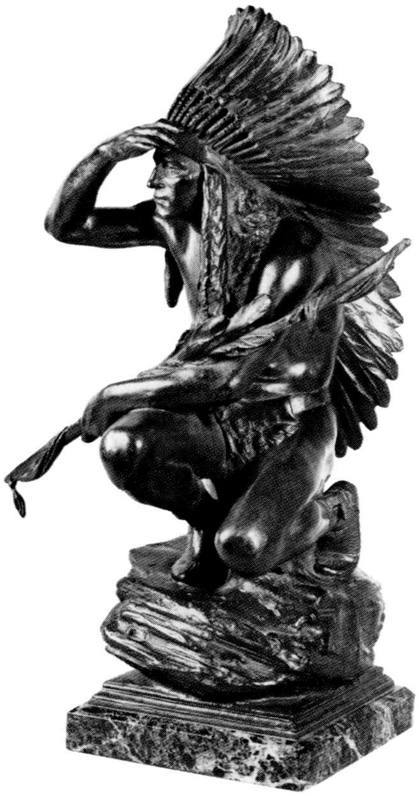

16.
On the Lookout
1919

Cast bronze, golden brown patina, H. 22½"; green marble base, 1½" x 8¼" x 9¼"

Signed and dated: *Massey Rhind. Sc/1919* (on rock at right)

Foundry mark: *Roman Bronze Works N-Y-* (on back of base)

Collection of The Newark Museum. Gift of Frederick W. Smith 1961
(61.30)

15.
Peter Stuyvesant
1913

Cast bronze, brown patina, H. 38¼"

Signed and dated: *J. Massey Rhind/1913* (on base at back right)

Foundry mark: *Roman Bronze Works N-Y-* (on base at back left)

Collection of The Newark Museum. Purchase 1914
(14.171)

This is a reduction of one of four ten-foot bronze sculptures the artist designed for the Exchange Court Building. The other figures were of Henry Hudson, General Wolfe and DeWitt Clinton. A larger cast of the Stuyvesant is in Bergen Square, Jersey City, New Jersey.

EDWIN WILLARD DEMING

Born 1860, Ashland, Ohio
Died 1942, New York City

Deming grew up on the prairie frontier in Illinois. His friendships with and interest in the life of the Indians of the area provided the bases for his work as both a painter and sculptor. The artist was largely self-taught, although he studied for a short time at the Art Students League in New York (1883) and the Académie Julian in Paris (1884-85). For most of his career, Deming lived and traveled in the West, painting the Indians he so admired. Sculpture forms a small portion of his artistic output and most date from the period, 1905-1910.

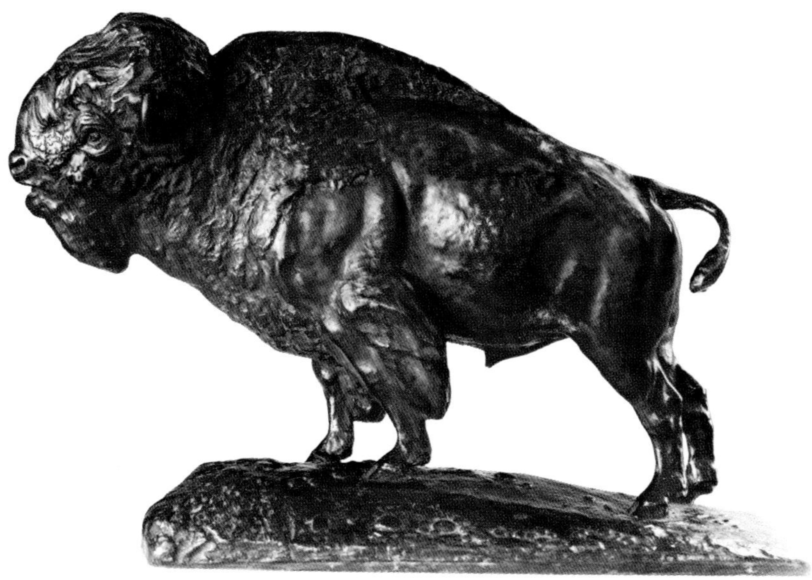

17.
Buffalo
1908

Cast bronze, dark brown patina, 14 9/16" x 17 5/16" x 7 1/16"

Signed and dated: *E W Deming C 1908* (on base)

Inscribed: *N° 6-* (underneath base)

Foundry mark: *Roman • Bronze • Works • N.Y.* (on base)

Lent by The Brooklyn Museum. Carll H. De Silver Fund

FREDERIC SACKRIDER REMINGTON
Born 1861, Canton, N.Y.
Died 1909, Ridgefield, Conn.

Active as a sculptor, illustrator, painter and writer, Remington is perhaps the best-known artist of the American West. His first formal training was at the Yale University School of Fine Arts, which he entered in 1878. After visiting the West in 1881, he lived briefly in Kansas and Missouri. Remington studied painting at the Art Students League with J. Alden Weir in 1886. Beginning the following year his illustrations of western life appeared regularly in publications such as *Harper's Weekly.* He made trips to the West almost yearly to collect fresh material for his work. At his home in New Rochelle, New York, Remington watched the sculptor Frederic W. Ruckstull working on his equestrian statue of *Maj. General John F. Hartranft* and, in 1894, decided to try his hand at the media. Remington's first sculpture, *The Bronco Buster* (1895), met with considerable acclaim and over the next fourteen yeas he enjoyed a successful career as a sculptor of cowboy, American Indian and cavalry subjects. Around 1900, Remington turned over the casting of his sculptures to the Roman Bronze Works in New York and was one of the first American artists to fully exploit the intricate surface textures to be derived from foundry's technique of lost-wax casting. The artist died suddenly of appendicitis shortly after moving to a new studio in Connecticut.

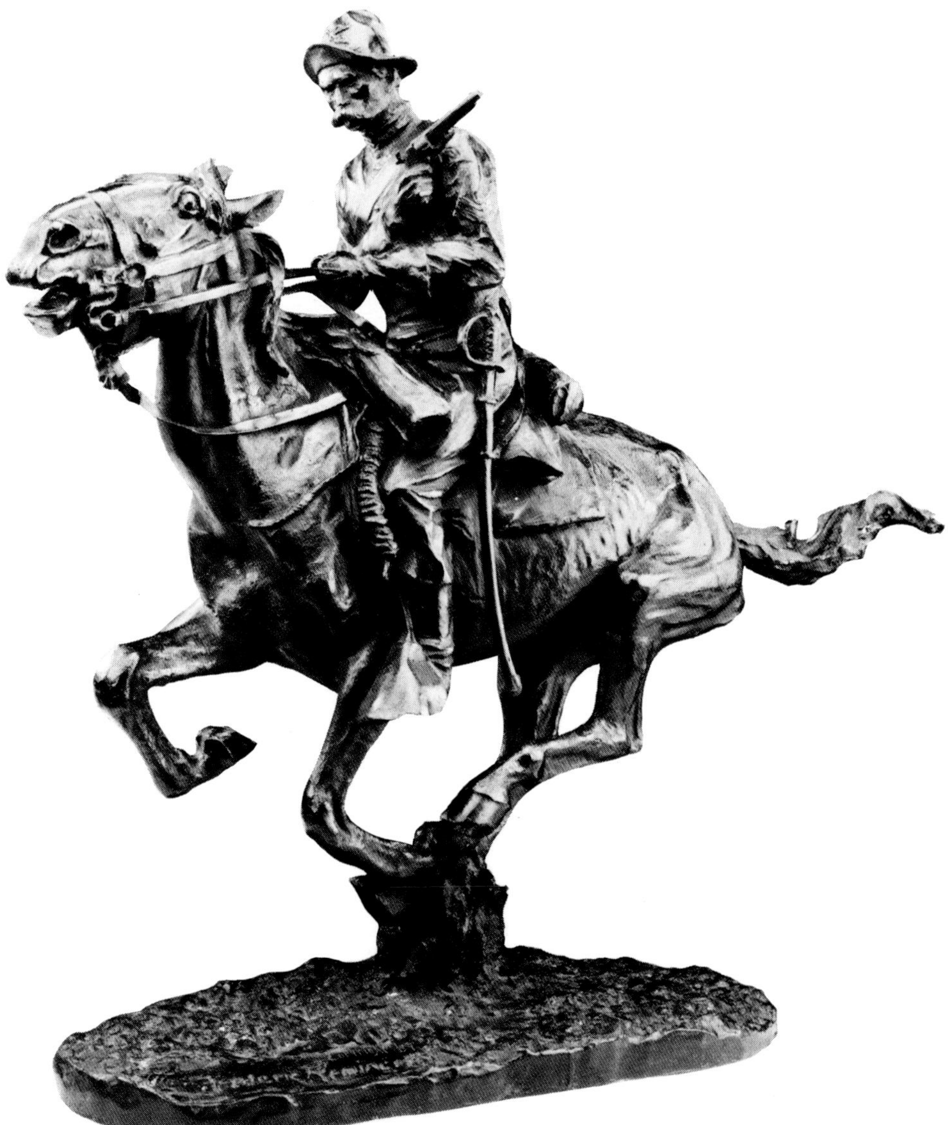

18.
Trooper of the Plains
1909; cast 1919

Cast bronze, dark green patina, 26¼″ x 30″ x 8½″

Signed: *Copyright by/Frederic Remington* (on base)

Inscribed: *N° 13* (underneath base)

Foundry mark: *Roman Bronze Works N-Y-* (on base at right)

Collection of The Newark Museum. Gift of Mr. and Mrs. William V. Griffin 1952 (52.167)

According to a ledger kept by the Roman Bronze Works, this was one of the last of fifteen castings made before the plaster was destroyed in 1919.

CHARLES SCHREYVOGEL
Born 1861, New York City
Died 1912, Hoboken, N.J.

After his family moved to New Jersey in the 1870s, Schreyvogel pursued his interest in art at the Newark Art League. H. August Schwabe, president of the school, was so impressed with the young artist's work that he helped finance his trip to Europe, where Schreyvogel studied painting in Munich with the American expatriate Carl Marr. Returning to the United States in the early 1890s, the artist established his home in Hoboken but often lived and traveled in the West for extended periods of time. Like Frederic Remington, Schreyvogel was active as both a painter and sculptor. He became a lifelong friend of Buffalo Bill Cody and many of the artist's subjects were the cowboys and Indians who performed in Cody's Wild West Show. Schreyvogel's first sculpture, *White Eagle* (1899), was modeled on one of his visits to the Ponca Indian Reservation in Oklahoma. Often, as with *The Last Drop*, the artist's sculptures were created as models for paintings and later cast in bronze.

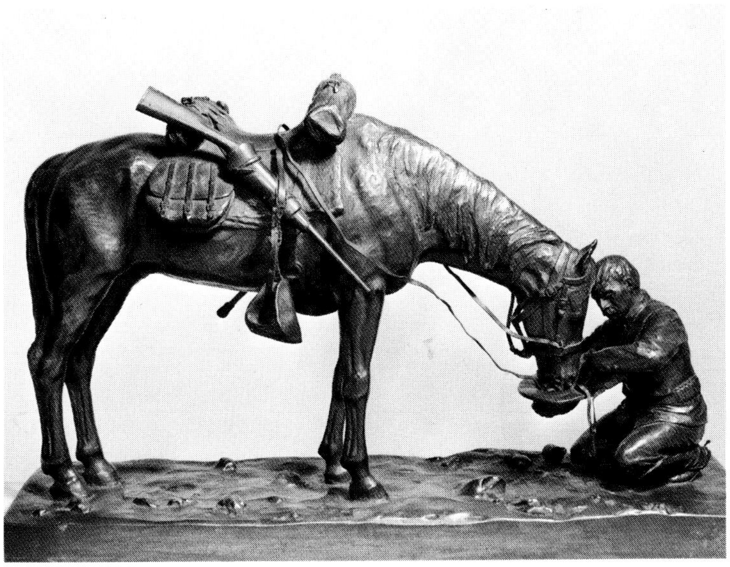

19.
The Last Drop
ca. 1900; cast 1904

Cast bronze, golden brown patina, 12″ x 18⅞″ x 5″

Signed and dated: *Copyrighted 1903 by/Chas Schreyvogel* (on base at back)

Foundry mark: *Roman Bronze Works N.Y. 1904* (on left edge of base)

Collection of The Newark Museum. Gift of Joseph S. Isidor 1925 (25.1280)

CYRUS EDWIN DALLIN
Born 1861, Springville, Utah
Died 1944, Arlington Heights, Mass.

While Remington and Schreyvogel usually depicted the adventure and violence of the American West, Dallin is best known for his dignified and sympathetic images of its Indians. Raised in the Utah wilderness, the artist moved east in 1880 and studied in Boston with the sculptor Truman Bartlett. By 1884, he had established his own studio, where he modeled portraits and small statues, including the first of his Indian subjects. In 1888, Dallin continued his studies with two years in Paris at the Académie Julian under Henri Chapu. His interest in Indian subjects was further stimulated when he saw Buffalo Bill's Wild West Show, which came to Paris in 1889. Upon returning to the United States the artist worked for several years in Salt Lake City (where he modeled an angel for the dome of the Mormon Temple) and taught at the Drexel Institute in Philadelphia. In 1897 he returned to Paris. During this second trip abroad Dallin studied with Jean Dampt and developed an enthusiastic European audience for his Indian subjects. In 1890, Dallin modeled the first of four large-scale equestrian sculptures—*The Signal of Peace, The Medicine Man, The Protest* and *The Appeal to the Great Spirit*—that were to become his most famous works and his most eloquent statements on the dignity of a defeated race. After returning to the United States in 1900, Dallin spent his career in Boston, where he taught for many years at the Massachusetts State Normal Art School and produced numerous public monuments.

20.
On the Warpath
1914-15

Cast bronze, brown patina, 9″ x 8″ x 2⅛″

Signed: *©C.E. Dallin* (on front of base)

Inscribed: *#31* (on edge of base)

Foundry mark: *Gorham Co. Founders 0489* (stamped on edge of base)

Collection of The Newark Museum. Anonymous Gift 1975 (75.192)

First exhibited at the Pennsylvania Academy of the Fine Arts, Philadelphia, in 1915, this work also exists in 33-inch and 23-inch versions.

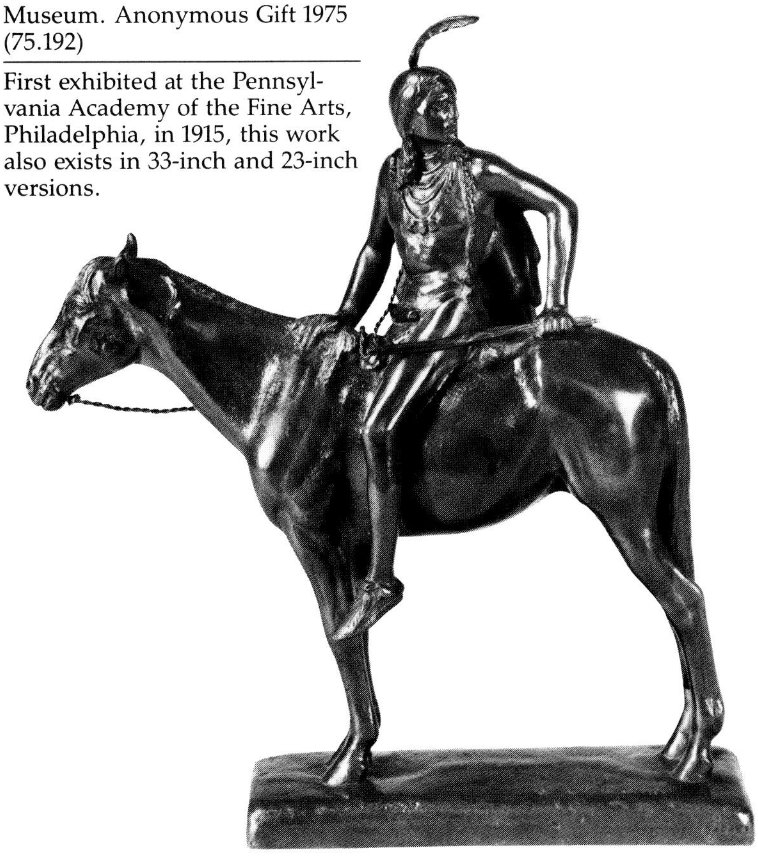

ALEXANDER PHIMISTER PROCTOR
Born 1862, Bosanquit, Ontario, Canada
Died 1950, Palo Alto, Calif.

Like other sculptors who are known primarily for Indian and animal subjects, Proctor spent his early life in the West. His first formal instruction was in 1888, when he moved to New York to study at the Art Students League and the National Academy of Design. Public recognition came in 1893 when he received a medal for his work at the Columbian Exposition in Chicago, which included more than thirty-five sculptures of animals of the American wilderness. The same year, he left for Paris and study in the ateliers of Denys Pierre Puech and Jean-Antoine Injalbert. After returning to New York, he assisted Saint-Gaudens in modeling the horses for his Logan and Sherman monuments and won the first Rinehart Scholarship, a travel fund established by the American Neoclassic sculptor William Henry Rinehart. This prize allowed Proctor to return to Paris for five years and work at the Académie Julian and the Académie Colarossi. His sculptures enjoyed great success in both America and Europe, where he was often compared to the French *animaliers* Antoine-Louis Barye and Emmanuel Frémiet. When he returned to the United States in 1900, the artist traveled widely and lived mostly in the western states, where many of his public monuments such as *Pioneer Mother* (1928, Kansas City, Missouri) are to be found. Two of his most familiar outdoor works are the monumental bisons on the Q Street Bridge in Washington, D.C., and the crouching tigers at the entrance of Nassau Hall at Princeton University.

21.
Fawn
1887-93

Cast bronze, brown patina,
6⅝″ x 7⅞″ x 2⅝″

Signed and dated:
A.P. Proctor/1893 (on base)

No foundry mark

Lent by The Brooklyn Museum.
Gift of George D. Pratt

This work was exhibited at the Exposition Universelle, Paris, 1900, where it won a gold medal.

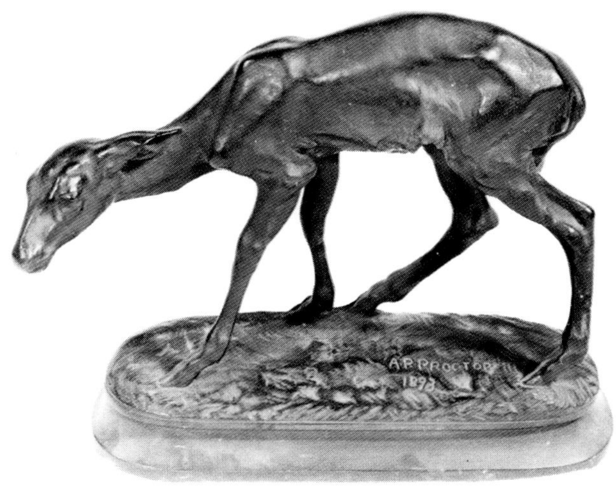

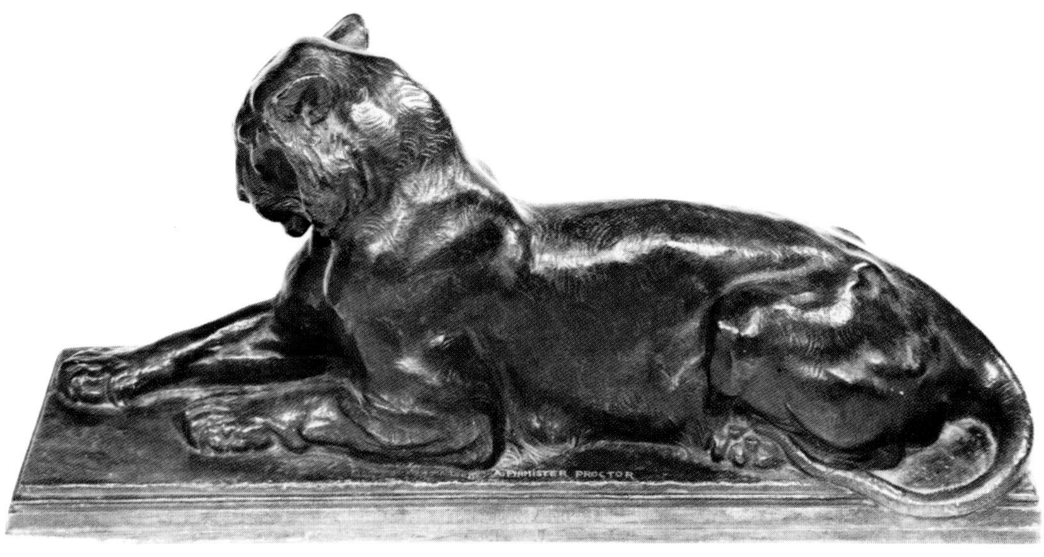

22.
Princeton Tiger
1908-09; cast 1912

Cast bronze, dark brown patina,
10″ x 22⅛″ x 6½″

Signed and dated: *A. Phimister Proctor © 1912* (on base)

No foundry mark

Lent by The Brooklyn Museum.
Gift of George D. Pratt

One of a pair of tigers commissioned by the Princeton Class of 1879 for the entrance to Nassau Hall and dedicated on Commencement Day 1909, this work won Proctor the gold medal awarded annually by the Architectural League of New York.

FREDERICK WILLIAM MacMONNIES

Born 1863, Brooklyn, N.Y.
Died 1937, New York City

With Olin L. Warner's tragic death in 1896, MacMonnies assumed his place in the triumvirate of leading sculptors that included Saint-Gaudens and French. MacMonnies began his career as an errand boy and studio assistant to Saint-Gaudens, while at the same time learning drawing at the National Academy of Design and the Art Students League. When the artist went to Paris to study in 1884, his earlier training enabled him to easily pass the entrance requirements for the Ecole de Beaux-Arts. After working briefly in Falguière's class, MacMonnies traveled to Munich where he studied painting and then, at Saint-Gaudens's request for help on several large projects, returned to New York. In 1889, he went back to France and remained there until shortly before World War I. Under the influence of Falguière and other French artists, his sculpture took on the lively, impressionistic modeling that led Taft to declare that "for sheer dexterity of manipulation there is no American sculptor to be compared with him." The straightforward naturalism of his nude figures was sometimes deemed too erotic or unwholesome for the American public. The Women's Christian Temperance Union was able to have the life-size version of *Bacchante and Infant Faun* removed from the court of the Boston Library. After it was accepted by the Trustees of the Metropolitan Museum of Art in the supposedly more liberal city of New York, the statue was greeted by petitions against it by the American Purity League and the Social Reform League. Publicity from this scandal, however, only seems to have brought the artist further public recognition. MacMonnies worked almost exclusively in bronze and among his best-known pieces are *Nathan Hale* (1893, City Hall Park, New York), the *Quadriga* (1898) on the Soldiers' and Sailors' Arch in Brooklyn, and *Washington Refusing Defeat at the Battle of Princeton* (1922, Princeton, New Jersey). MacMonnies was married to the painter Mary Fairchild and spent much of his later career painting.

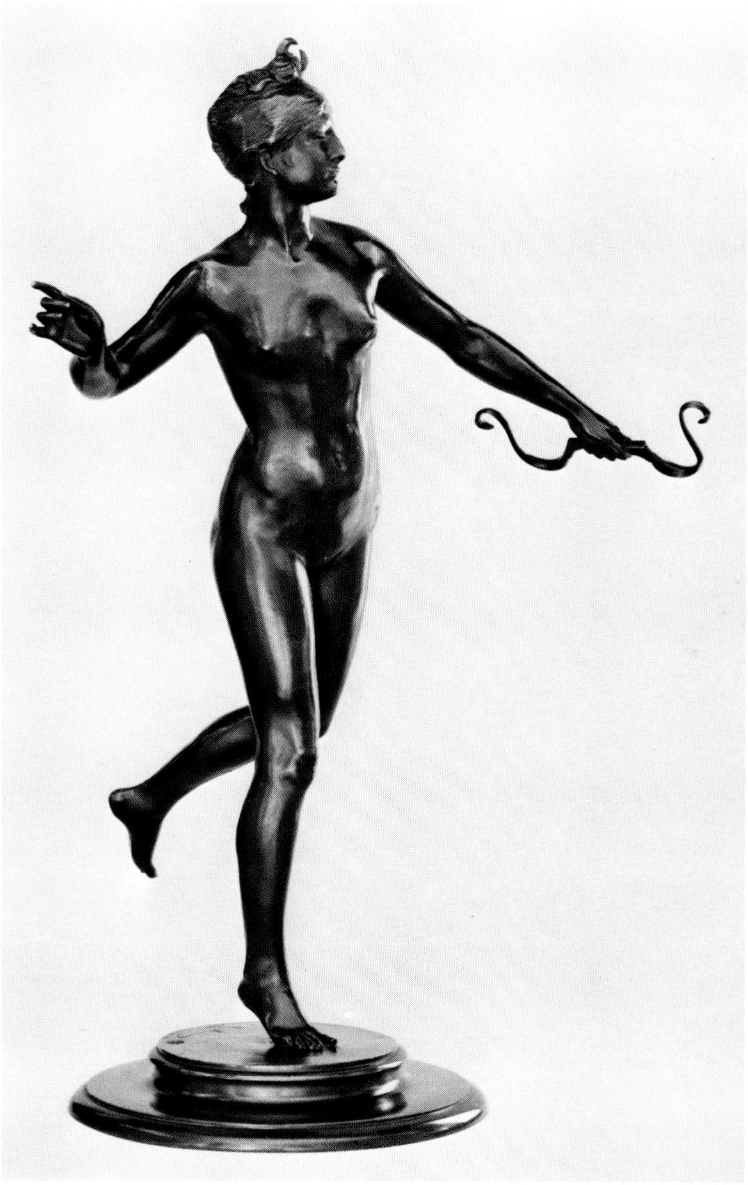

23.
Diana
1889-90

Cast bronze, dark brown patina, H. 30¾″

Signed and dated: *F. MacMonnies 1890* (on base at back)

Foundry mark: *E. Gruet/Fondeur/Paris* (stamped on top of base at back)

Inscribed: *R.6.* (underneath base)

Lent by Jordan-Volpe Gallery, New York

The plaster for this sculpture was exhibited at the Paris Salon of 1889, where it won an honorable mention.

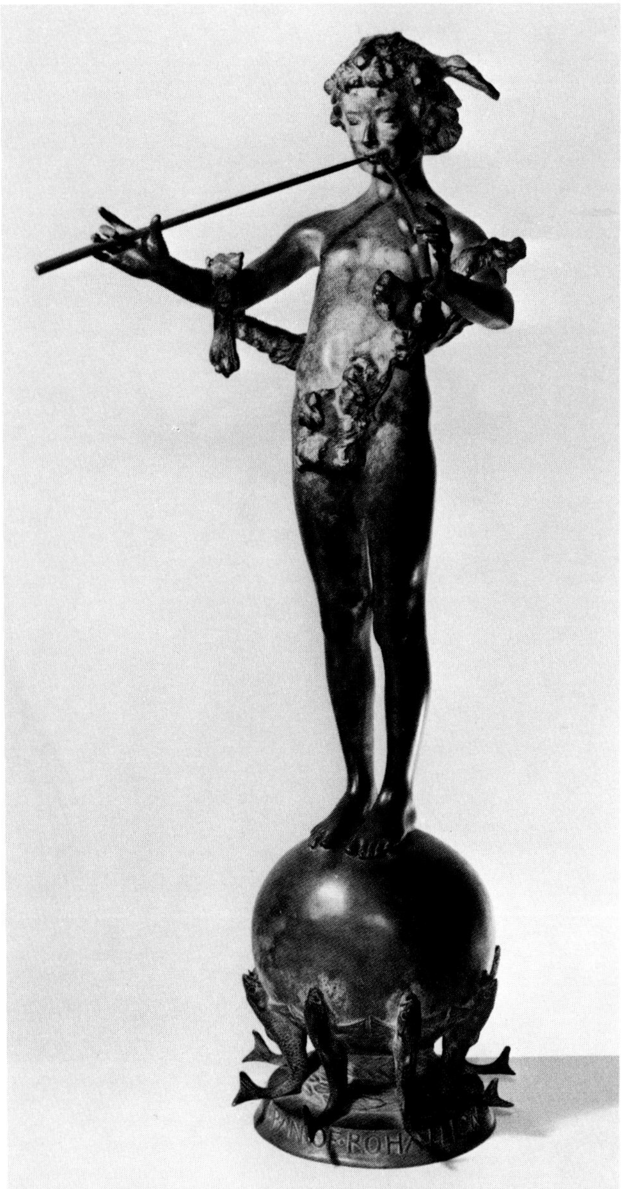

24.
Pan of Rohallion
1890

Cast bronze, green patina, H. 30″

Signed and dated: *France/ Frederick MacMonnies/Copyright 1894/Paris 1890* (on back beneath figure's feet)

Inscribed: *To • Pan • of • Rohallion • Anno Domini M • D • C • C • C • LX • L* (around bottom of base)

Foundry mark: *H. Rouard, Fondeur* (on back beneath signature)

Collection of The Newark Museum. Gift of Mr. and Mrs. Felix Fuld 1926 (26.2365)

This piece was designed for the grounds of "Rohallion," the estate of Edward D. Adams in Rumson, New Jersey. The original was over life-size. Adams gave the artist permission to make unlimited replicas not to exceed 36 inches high.

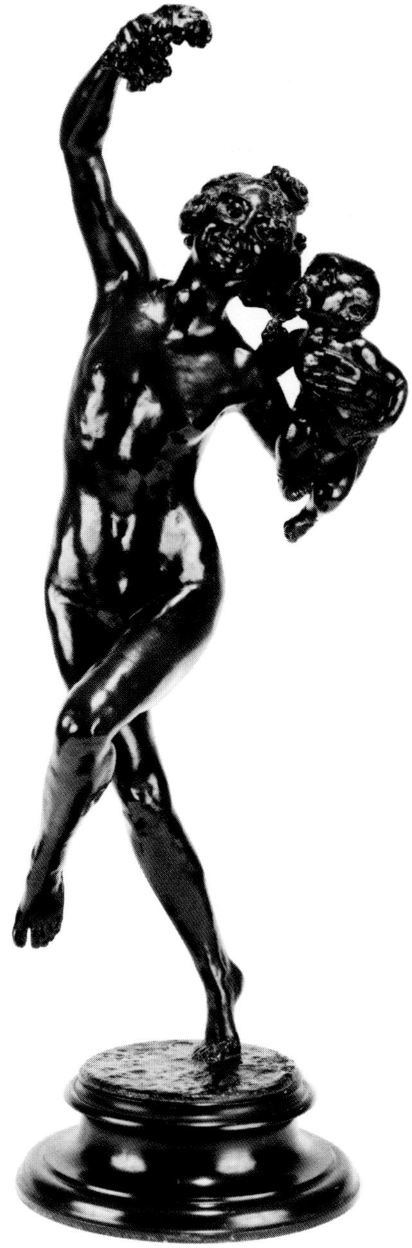

25.
Bacchante and Infant Faun
1894

Cast bronze, dark brown patina, 34″ x 9″ x 12″

Signed and dated: *F. MacMonnies 1894* (on top of base)

Foundry mark: *Jabouff et Rouard Fondeurs Paris* (stamped on top of base)

Lent by Conner-Rosenkranz, New York

26.
Infant Faun
Not dated (probably ca. 1890-95)

Cast bronze, dark green patina, 51½" x 7⅛" x 5¾"

Signed: *Frederick MacMonnies* (on back)

Foundry mark: *R.B.* [Roman Bronze] *4* (underneath)

Lent by Robert Mehlman, New York

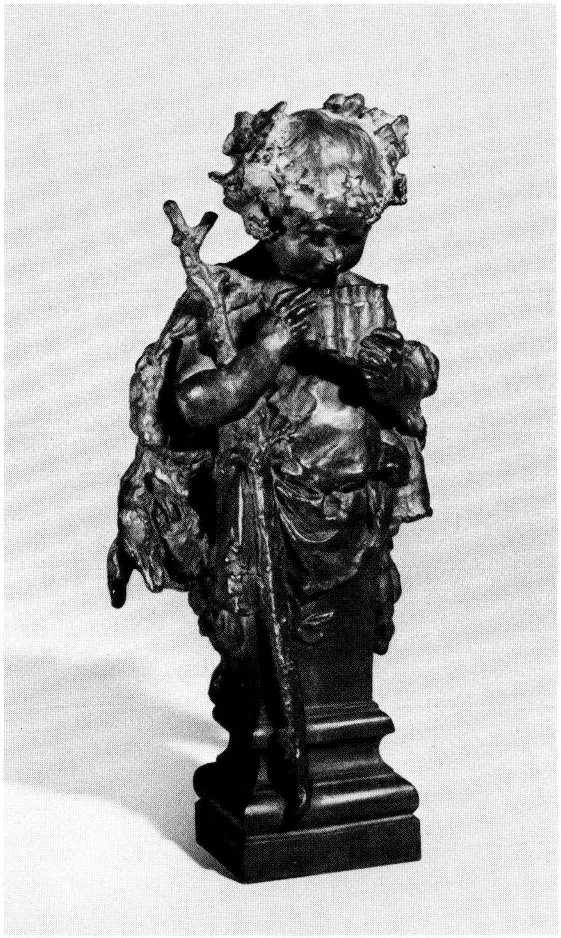

JOHN FLANAGAN
Born 1865, Newark, N.J.
Died 1952, New York City

As a young man, Flanagan studied at the Cooper Union in New York and was employed by the Perth Amboy Terra Cotta Works in New Jersey. He then worked for three years in Saint-Gaudens' studio, while attending night classes at the Art Students League with George de Forest Brush. In 1890 he was able to go to Paris, where he remained until 1902. There, he studied in several art schools, including Falguière's atelier at the Ecole des Beaux-Arts, and assisted MacMonnies on his monumental *Barge of State* for the Columbian fountain of the 1893 Chicago exposition. Although Flanagan is known as one of America's most talented medalists, he also executed a number of public sculptures, including the clock in the Rotunda Reading Room of the Library of Congress and a bronze relief for the Newark Free Public Library.

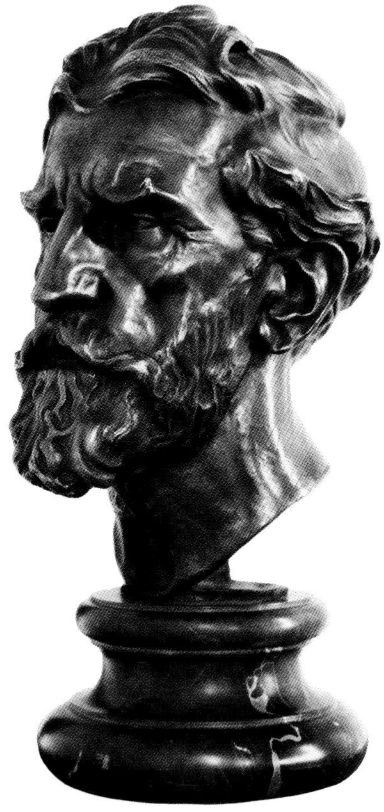

27.
Augustus Saint-Gaudens
1904-24

Cast bronze, dark green patina, H. 16¼"; circular black marble base, H. 5"

Signed and dated: *John Flanagan 1924* (on back of neck)

Foundry mark: *Kunst - Foundry - N.Y.* (on pl back of neck)

Collection of The Newark Museum. Gift of Mr. and Mrs. Felix Fuld 1926 (26.2361)

The artist stopped work on this bust after Saint-Gaudens' death in 1907. It was resumed in 1920 and the plaster completed in 1924.

HERMON ATKINS MacNEIL
Born 1866, Chelsea, Mass.
Died 1947, New York City

MacNeil studied at the Massachusetts Normal Art School in Boston and then taught modeling for three years at Cornell University. In 1888, he went to Paris for further training and, like many others of his generation, chose the ateliers of Falguière and Chapu. Immediately upon his return to the United States in 1891, MacNeil moved to Chicago to work as an assistant on the sculptural decorations for the World's Columbian Exposition. The artist was fascinated by the Indians who came to participate in the fair and a performer from Buffalo Bill's Wild West Show served as model for the first of his numerous American Indian subjects. In 1896 MacNeil shared the Rinehart Scholarship with A.P. Proctor and worked in Rome for the next three years. After returning to New York, the artist received numerous public commissions for monuments, many of Indian subjects, and contributed substantially to the expositions held in Buffalo, Saint Louis and San Francisco. His interest in Indian subjects waned after 1910. MacNeil served as president of the National Sculpture Society and designed several medals and the United States quarter dollar (1916). For many years his studio was in College Point, Long Island, where a park is named after the artist.

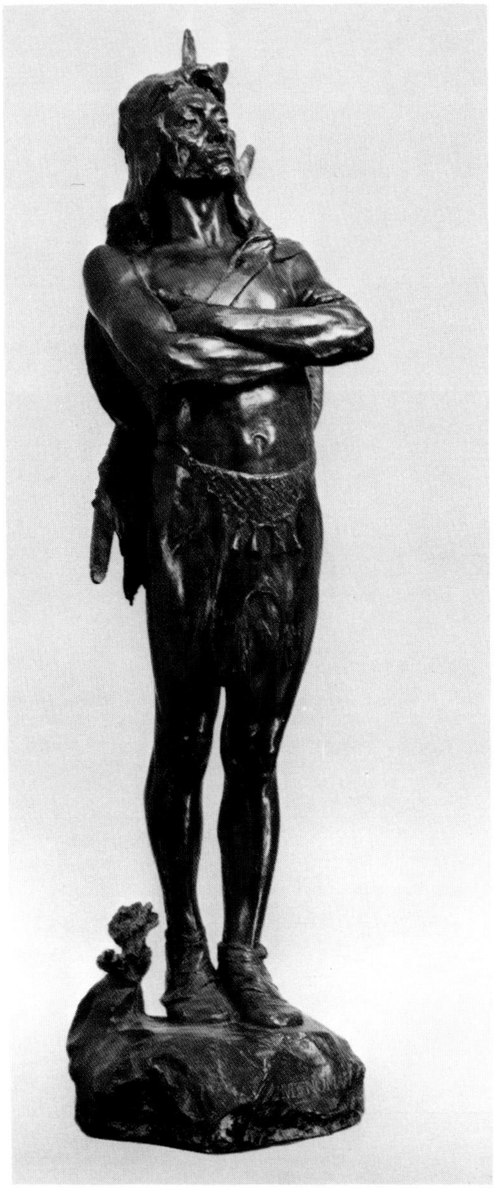

28.
A Chief of the Multnomah Tribe
1905

Cast bronze, dark brown patina, 37½" x 12" x 10"

Signed and dated: *H.A. MacNeil/05.* (on base at back)

Foundry mark: *Roman Bronze Works N.Y.* (on base at back)

Inscribed: *MVLTNOMAH* (on front of base) and *No 13* (underneath base)

Lent by the National Academy of Design, New York

This is a reduction of one of the two figures from the monument *The Coming of the White Man* (1905, Washington Park, Portland, Oregon). The Chief and his companion, the Medicine Man, represent members of the Multnomah Tribe of Oregon.

KARL THEODORE FRANCIS BITTER
Born 1867, Rudolfsheim, Austria
Died 1915, New York City

Bitter arrived in New York in 1889 virtually penniless and unable to speak English. His previous training at the School of Applied Arts and the Academy of Fine Arts in Vienna gave him the skills necessary to find employment with a New York firm that specialized in architectural decoration. His work soon came to the attention of the prominent architect Richard Morris Hunt, who adopted Bitter as his protégé. Throughout his career the artist received prestigious architectural commissions, such as a pair of bronze doors for Trinity Church in New York, stone figures for the facades of the Metropolitan Museum of Art and The Brooklyn Museum, and terra-cotta pediment figures for Pennsylvania Station in Philadelphia. In 1896, Bitter built a large studio in Weehawken, New Jersey, and employed many assistants to meet the demand for his work. He was appointed director of sculpture for the Buffalo, Saint Louis and San Francisco expositions. In addition to his architectural work, Bitter executed a number of public monuments, the most famous of which is the figure of Pomona, or "Abundance," (1915) for the Pulitzer Fountain at Fifth Avenue and Central Park South in New York. The artist's career ended tragically when he was struck by an automobile while returning with his wife from an evening at the opera.

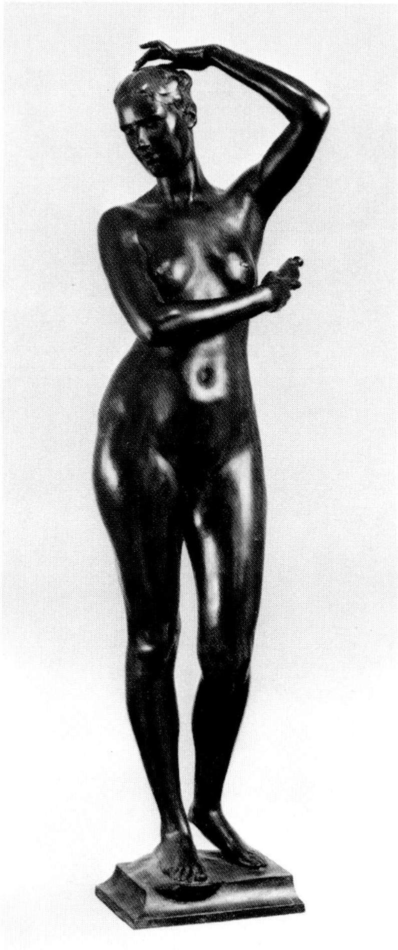

29.
Venus
Not dated (probably ca. 1900)

Cast bronze, 24¾" x 6" x 4"

Foundry mark: *Bureau Bros./ Phila.* (stamped underneath base)

Lent by the National Academy of Design, New York

ROBERT TAIT McKENZIE
Born 1867, Almonte, Ontario, Canada
Died 1938, Philadelphia, Pa.

McKenzie received a degree in medicine from McGill University in Montreal in 1892 and was appointed medical director of physical education and demonstrator in anatomy at his alma mater. He moved to the United States in 1904 to become director of the department of physical education at the University of Pennsylvania, a post he held until 1930. He then became research professor of physical education, writing many books on the subject. It was McKenzie's interest in anatomy that led him to begin sculpting in 1902. The vast majority of his sculptures are of athletes engaged in various sports and are located in a large collection of his works at Yale University. Although entirely self-taught and working outside official art circles, McKenzie also modeled portraits and a few public monuments.

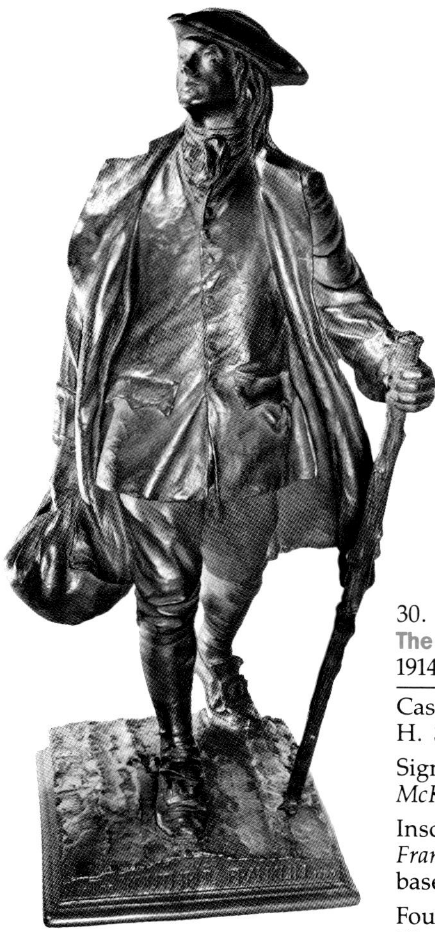

30.
The Youthful Franklin
1914

Cast bronze, brown patina, H. 36½"

Signed and dated: *R. Tait McKenzie/1914*

Inscribed: *1723/1706 The Youthful Franklin 1790* (across front of base)

Foundry mark: *Roman Bronze Works N-Y-*

Collection of The Newark Museum. Purchase 1914 (14.1073)

The life-size version of this figure was modeled for the University of Pennsylvania in Philadelphia, where it stands in front of Weightman Hall.

JOHN GUTZON De La MOTHE BORGLUM

Born 1867, near Bear Lake, Idaho
Died 1941, Chicago, Ill.

The son of Danish pioneers, Borglum was raised in the West. His first training was as a lithographer's apprentice in Los Angeles and at the Mark Hopkins Art Institute in San Francisco with William Keith and Virgil Williams. About 1890 he went to Paris for further study, working at the Académie Julian, Ecole des Beaux-Arts, and in individual studios. He was greatly influenced by the sculpture of Auguste Rodin and Jules Dalou. Although he first exhibited as a painter, his small bronze *Death of the Chief* was shown at the 1891 exhibition of the Société Nationale des Beaux-Arts and won him membership in the organization. After a short return to the United States, Borglum worked in England from 1896 to 1902. There he found great success as a painter and sculptor, enjoying the patronage of members of the English nobility. In 1902, Borglum returned to the United States and settled in New York City. His major commissions include the statues of the Twelve Apostles for the Cathedral of St. John the Divine, New York (1905), a head of Abraham Lincoln for the Library of Congress (1908) and the portraits of Washington, Jefferson, Lincoln and Roosevelt carved into Mount Rushmore, South Dakota (begun 1927). Borglum was a prolific artist and completed as many as 170 statues and monuments in his lifetime. During the 1910s and 1920s, his home in Stamford, Connecticut, was a popular summer retreat for artist friends such as George Luks.

31.
The Mares of Diomedes
1904

Cast bronze, green-brown mottled patina, 21″ x 34″ x 6″

Signed and dated: *Gutzon Borglum/1904* (on base)

No foundry mark

Collection of The Newark Museum. Gift of Joseph G. Spurr 1916 (16.568)

The original subject of this sculpture was an Indian stealing horses. The classical title, alluding to Hercules' capture of the man-eating horses of Thrace, was added later. Borglum stated of this piece, "The name is a convenience—the motive of the group, mainly intense controlled action." The large version (H. 62″) won a gold medal at the Louisiana Purchase Exposition in Saint Louis (1904) and is now in the collection of The Metropolitan Museum of Art in New York.

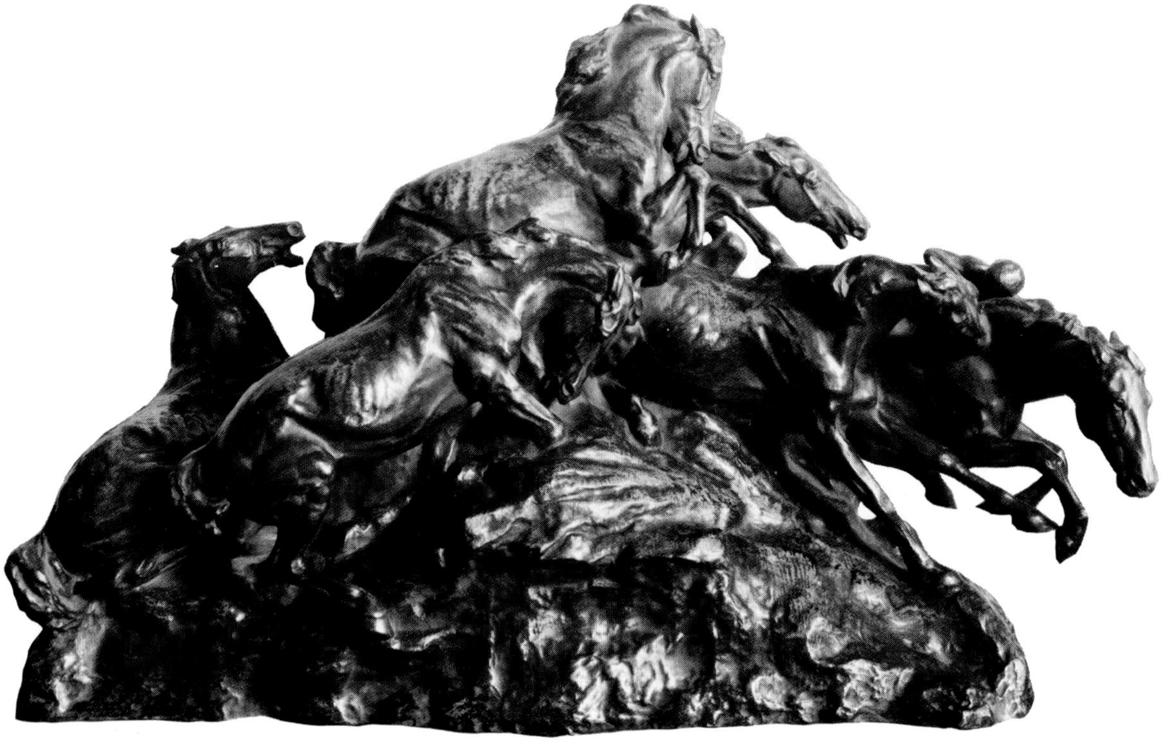

SOLON HANNIBAL BORGLUM
Born 1868, Ogden, Utah
Died 1922, Stamford, Conn.

Borglum managed his father's cattle ranch in Nebraska until his older brother Gutzon encouraged him to pursue his talent as an artist. In 1895 he moved to Cincinnati and studied for two years at the Art School under the sculptor Louis Rebisso. After winning local prizes, Borglum went to Paris in 1897. There, he was discouraged by his course of instruction at the Académie Julian and spent most of his time working independently and visiting one of the city's stables. After a brief return home, Borglum settled in New York in 1901. In 1906 he bought a farm in Silvermine, Connecticut, where he converted a barn for his studio. Until late in his career, the artist worked almost entirely with Western subjects and created a number of equestrian monuments. He enjoyed teaching and after returning from service in World War I, opened the American School of Sculpture in New York. His book on anatomy, *Sound Construction*, was published posthumously in 1923.

32.
Seated Lincoln
1909-11

Cast bronze, brown patina, 22″ x 23¾″ x 16¾″

Signed: *Gutzon Borglum*

Foundry mark: *Gorham Co Founders Q 501*

Collection of The Newark Museum. Purchase 1978 The Members' Fund (78.89)

Borglum began work on his commission for a seated Lincoln to be placed in front of the Essex County Court House, Newark, in 1909. It was cast by Gorham Co. in 1911 and unveiled on Memorial Day of that year.

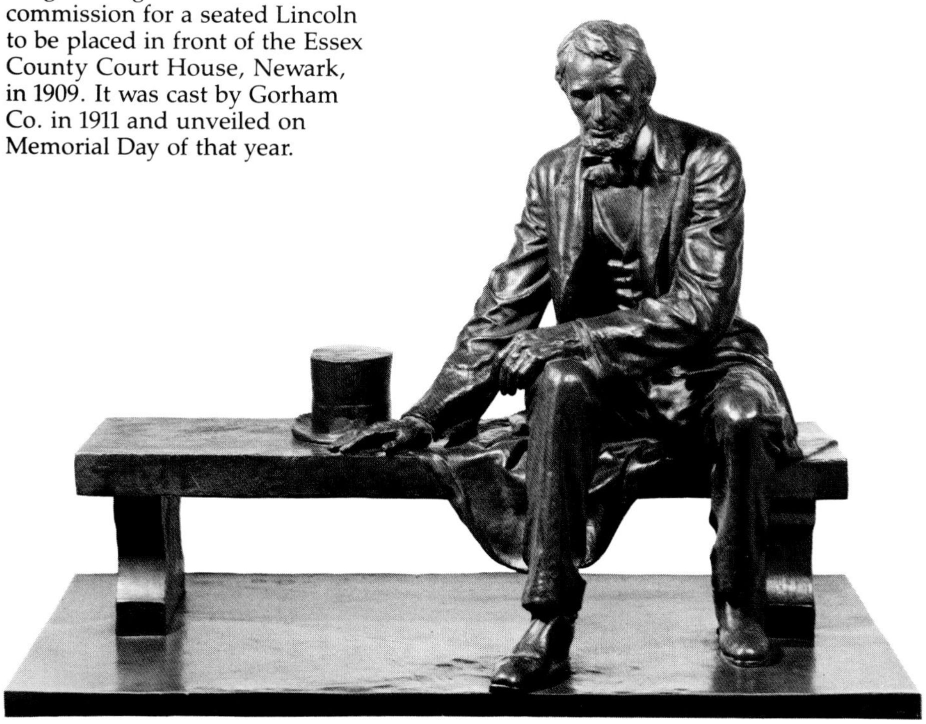

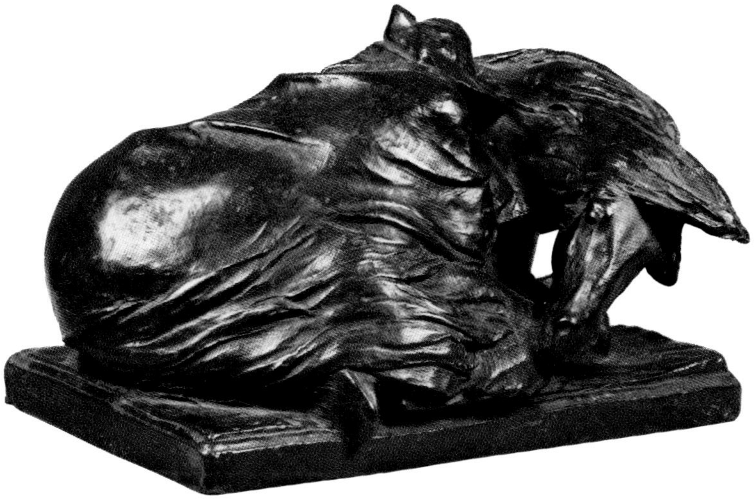

33.
Blizzard
ca. 1900

Cast bronze, green-brown patina, 6¼″ x 10⅜″ x 7¾″

Signed: *Solon Borglum* (on top of base near front)

Inscribed: *Solon Borglum/ Copyrighted* (on left edge of base) and *N.1.* (underneath)

No foundry mark

Collection of The Newark Museum. Gift of Mrs. J.G. Phelps Stokes 1977 (77.21)

ADOLPH ALEXANDER WEINMAN

Born 1870, Karlsruhe, Germany
Died 1952, Port Chester, N.Y.

Weinman was brought to the United States by his widowed mother and, at the age of fifteen, apprenticed to a New York wood and ivory carver. He also attended classes at Cooper Union and the Art Students League, where Saint-Gaudens encouraged his work. Much of his early career was spent as an assistant to other sculptors, including Warner, French, Saint-Gaudens, Charles Niehaus and Philip Martiny. In 1902 he opened his own studio. The success of a sculpture group made for the 1904 Saint Louis exposition, *The Destiny of the Red Man,* led to numerous commissions for architectural decorations and monuments. The naturalism of Weinman's early style was replaced in the 1920s and 1930s by a more decorative classicism. In 1916, Weinman designed the United States dime and half dollar.

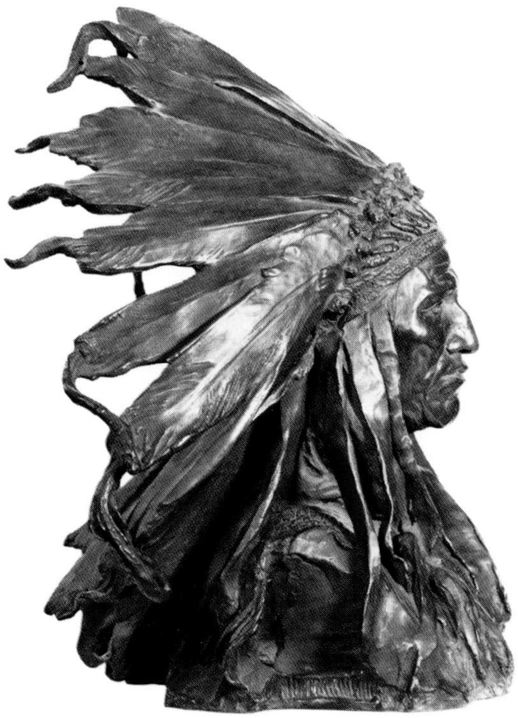

34.
Chief Blackbird, The Ogalla Sioux
1903; cast 1907

Cast bronze, brown patina, 16⅜″ x 12½″ x 11½″

Signed and inscribed: *Chief/Black/Bird/Ogalla/Sioux/A.A. Weinman* (on pl shoulder near bottom)

Foundry mark: *Roman Bronze/Works N.Y.* (on pr shoulder near bottom)

Lent by The Brooklyn Museum. Gift of George D. Pratt

This work is related to the full-length figure of Chief Blackbird used in *The Destiny of the Red Man* (1904).

35.
The Duet
1924

Cast bronze, dark brown patina, 23¼″ x 25″ x 16″

Signed: *A.A. Weinman Fecit* © (on base at right)

Inscribed: *N° 7* (underneath base)

Foundry mark: *Roman Bronze Works N-Y-* (on base at back)

Lent by Conner-Rosenkranz, New York, and Jordan-Volpe Gallery, New York

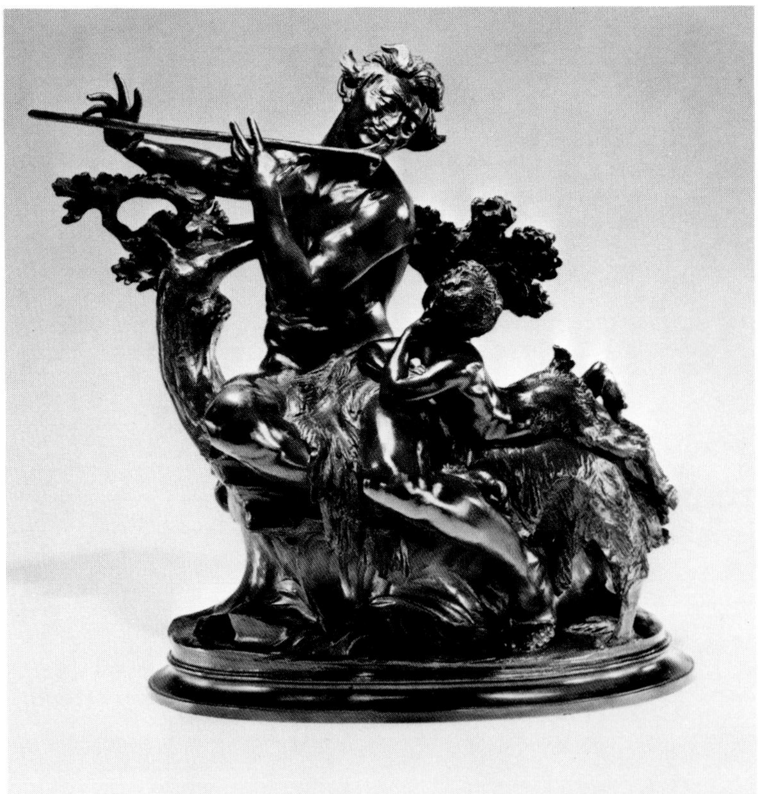

EDITH WOODMAN BURROUGHS
Born 1871, Riverdale-on-Hudson, N.Y.
Died 1916, Flushing, N.Y.

At the age of fifteen, Woodman entered the drawing class of Kenyon Cox and the modeling class of Saint-Gaudens at the Art Students League. Early in her career she became successful as a designer of fountains and small decorative figures, a specialty that continued in later years. In 1893, Woodman married the painter Bryson Burroughs, later curator of painting at The Metropolitan Museum of Art. Together they spent two years studying in Paris, where she received instruction from Injalbert and the painter Luc-Olivier Merson. The couple returned to New York City in 1905, both pursuing their careers. While visiting Paris in 1909, the artist discovered the work of the French sculptor Auguste Maillol and the lively naturalism of her earlier work gave way to simpler, more monumental forms. Mrs. Burroughs died only a year after completing her most ambitious works, two large fountains for the Panama-Pacific Exposition in San Francisco (1915).

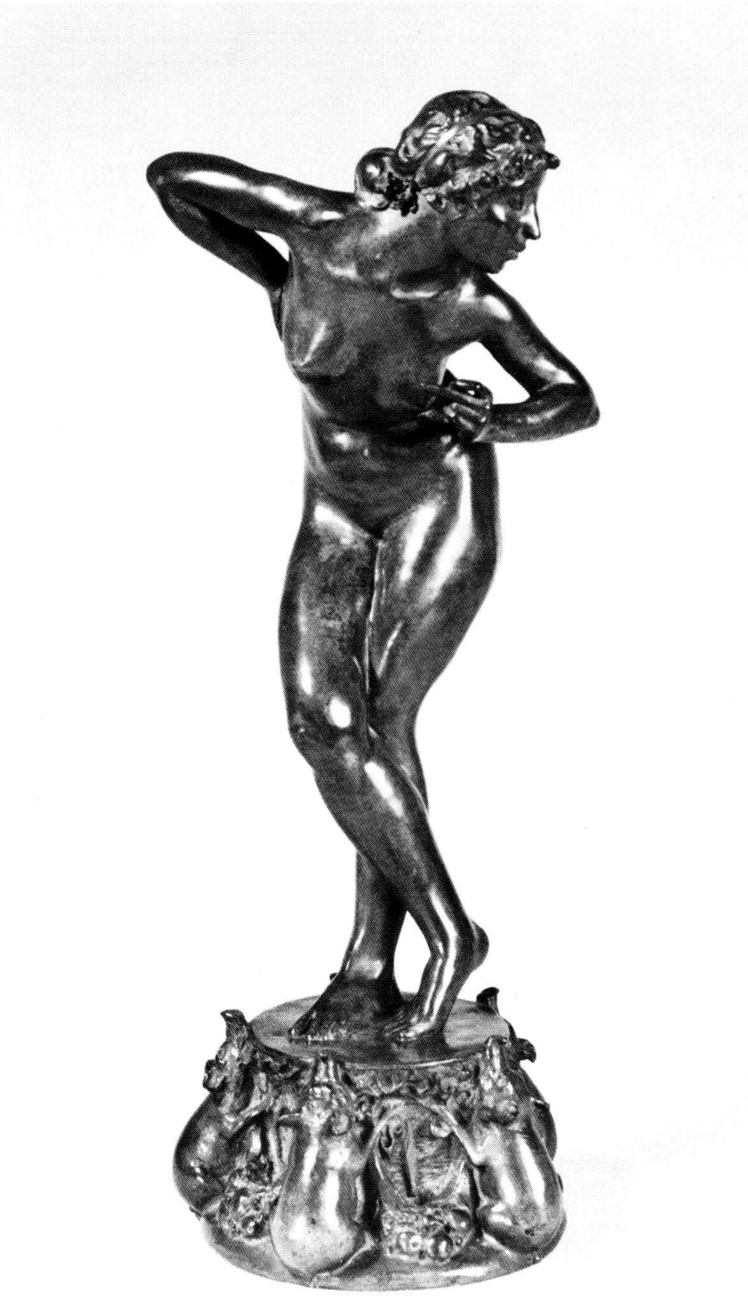

36.
Circe
1907

Cast bronze, green patina, 20½″ x 9″ x 7½″

Signed: *Edith Woodman Burroughs* (on base at back)

Inscribed: *Copyright 1907 3rd cast* (on base at back, near signature)

Collection of The Newark Museum. Purchase 1914 (14.18)

BESSIE ONAHOTEMA POTTER VONNOH
Born 1872, St. Louis, Mo.
Died 1955, New York City

As with Edith Woodman Burroughs, Bessie Potter's career has been closely associated with that of an artist husband. As a young woman, she studied at the Art Institute of Chicago with Lorado Taft and assisted him with sculptural decorations for the World's Columbian Exposition. Potter made several trips to Europe, but never studied in a particular school or atelier. Her sculpture was influenced by the impressionistic modeling of contemporary French artists, especially Paul Troubeskoy whose small figure and portrait studies she had seen at the Chicago fair. Potter married the painter Robert Vonnoh in 1899. Mrs. Vonnoh became famous for her sentimental rendering of women and children, the latter of which she often incorporated into fountain designs. By the time her career came to a close in the 1930s, she had been honored with a solo exhibition at The Brooklyn Museum (1913) and had won numerous awards.

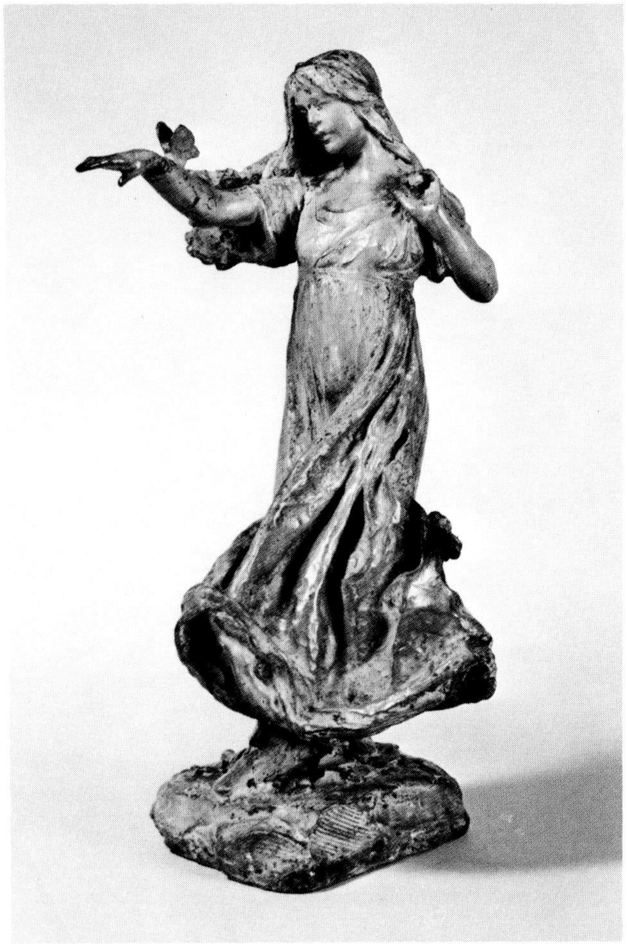

37.
Girl with Butterflies
Not dated (probably ca. 1900)

Cast bronze, bright green patina, 13¼" x 6½" x 6½"

Signed and numbered: *Bessie Potter Vonnoh/no. VI.* (on base at back)

Foundry mark: *Roman Bronze Works N.Y.* (on base at left)

Collection of The Newark Museum. Gift of Joseph S. Isidor 1914
(14.408)

ANNA VAUGHN HYATT HUNTINGTON
Born 1876, Cambridge, Mass.
Died 1973, Redding, Conn.

Of the many women sculptors who began their careers in the early 1900s, few enjoyed the success of Anna Hyatt. Born the daughter of a Harvard professor of paleontology, she moved to New York around 1900 and studied at the Art Students League with Hermon MacNeil. At about the same time, the artist also received criticism from Gutzon Borglum whose equestrian sculptures influenced her greatly. Like Bessie Potter Vonnoh, Hyatt became famous for a particular genre of sculpture: animals and equestrian monuments. Although working in the tradition of the nineteenth-century French *animaliers,* her figures are more broadly modeled than those of Barye or Fremiet. Two of her best-known monuments are those of *Joan of Arc* (1915, Riverside Drive, New York City) and *Cid Campeador* (1927), a replica of which is on the grounds of the Hispanic Society, New York City. In 1923 she married the railroad magnate Archer M. Huntington. The artist continued her career as a sculptor until quite late in life. Her statue of the Cuban patriot José Martí riding into battle was erected at the Sixth Avenue entrance to Central Park in 1965. Mrs. Huntington and her husband established Brookgreen Gardens in South Carolina, which houses in an outdoor setting one of the most complete collections of American sculpture from the late nineteenth century through the 1940s.

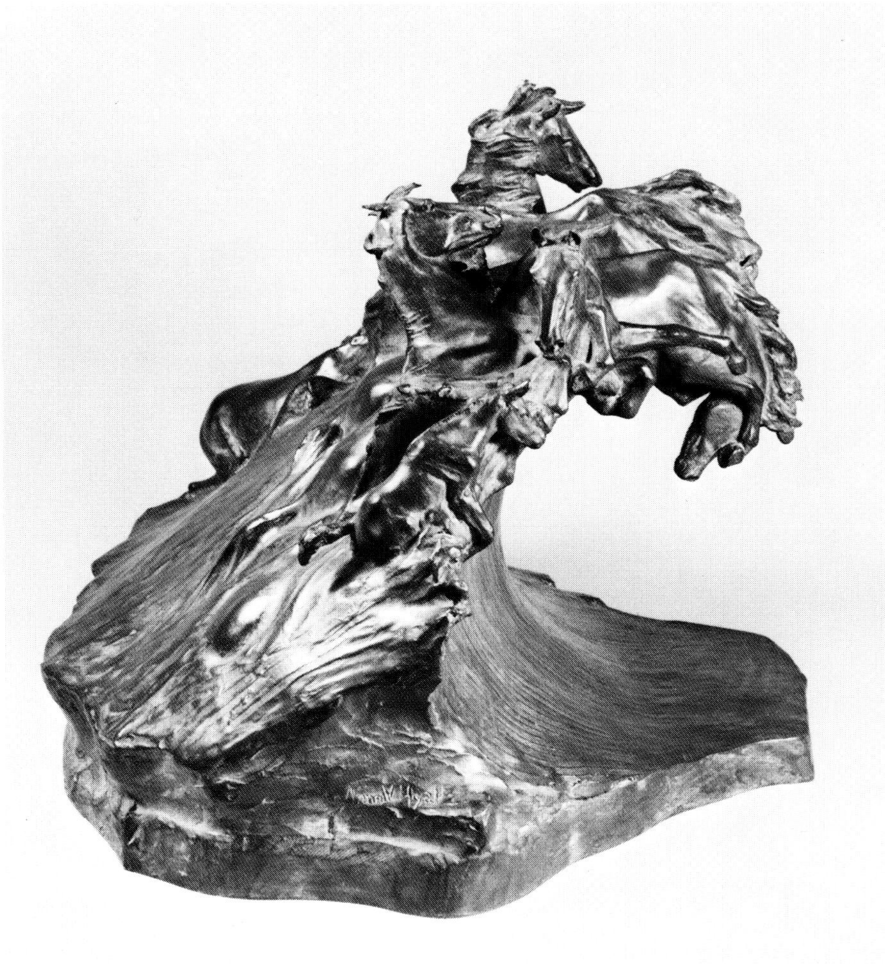

38.
Riders to the Sea
ca. 1912

Cast bronze, dark green patina, 18½″ x 24″ x 21″

Signed: *Anna V. Hyatt* (on base at front)

Foundry mark: *Roman Bronze Works N.Y.* (on base at back)

Collection of The Newark Museum. Gift of the Estate of Mrs. Florence P. Eagleton 1954 (54.135)

39.
Yawning Tiger
Not dated

Cast bronze, brown patina, 8¾″ x 28″ x 5″

Signed: *Anna V Hyatt* (at bottom of pr leg)

Inscribed: *#56* (at bottom of pl leg)

Foundry mark: *Gorham Co. Founders/0509* (stamped at bottom of pr leg)

Collection of The Newark Museum. Gift of Mr. and Mrs. Felix Fuld 1926 (26.2364)

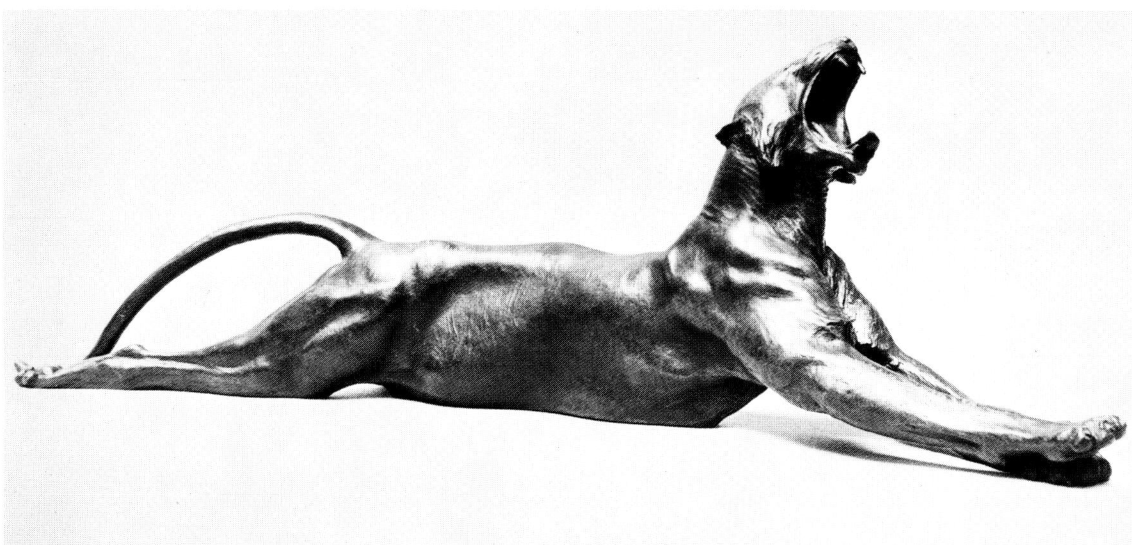

HERBERT HASELTINE
Born 1877, Rome, Italy
Died 1962, Paris, France

The son of the American landscape painter William Stanley Haseltine, the sculptor was raised in Italy and educated at Harvard College. With the encouragement of his family he studied art at the Royal Academy in Munich and at the Académie Julian in Paris. His first attempts at sculpture came when his teacher Aimé Morot suggested he try modeling to understand the forms in his paintings. One of Haseltine's first efforts as a sculptor, a group of polo players exhibited at the Salon of 1906, was so successful that he spent the rest of his career modeling animals, especially thoroughbred horses. Around 1920, after service in World War I, the artist became greatly influenced by Egyptian sculpture in the Louvre. This influence gave his works a greater heaviness and simplification of form and also seemed to have sparked his interest in adding color and stone inlay. Between 1921 and 1924, he executed a group of twenty marble sculptures of British champion animals, a complete set of which are in the Field Museum, Chicago. They proved so successful that he cast versions of these in bronze and exhibited them widely in America and Europe. Haseltine also was commissioned to do several large-scale statues, including the equestrian *George Washington* in front of the National Cathedral in Washington.

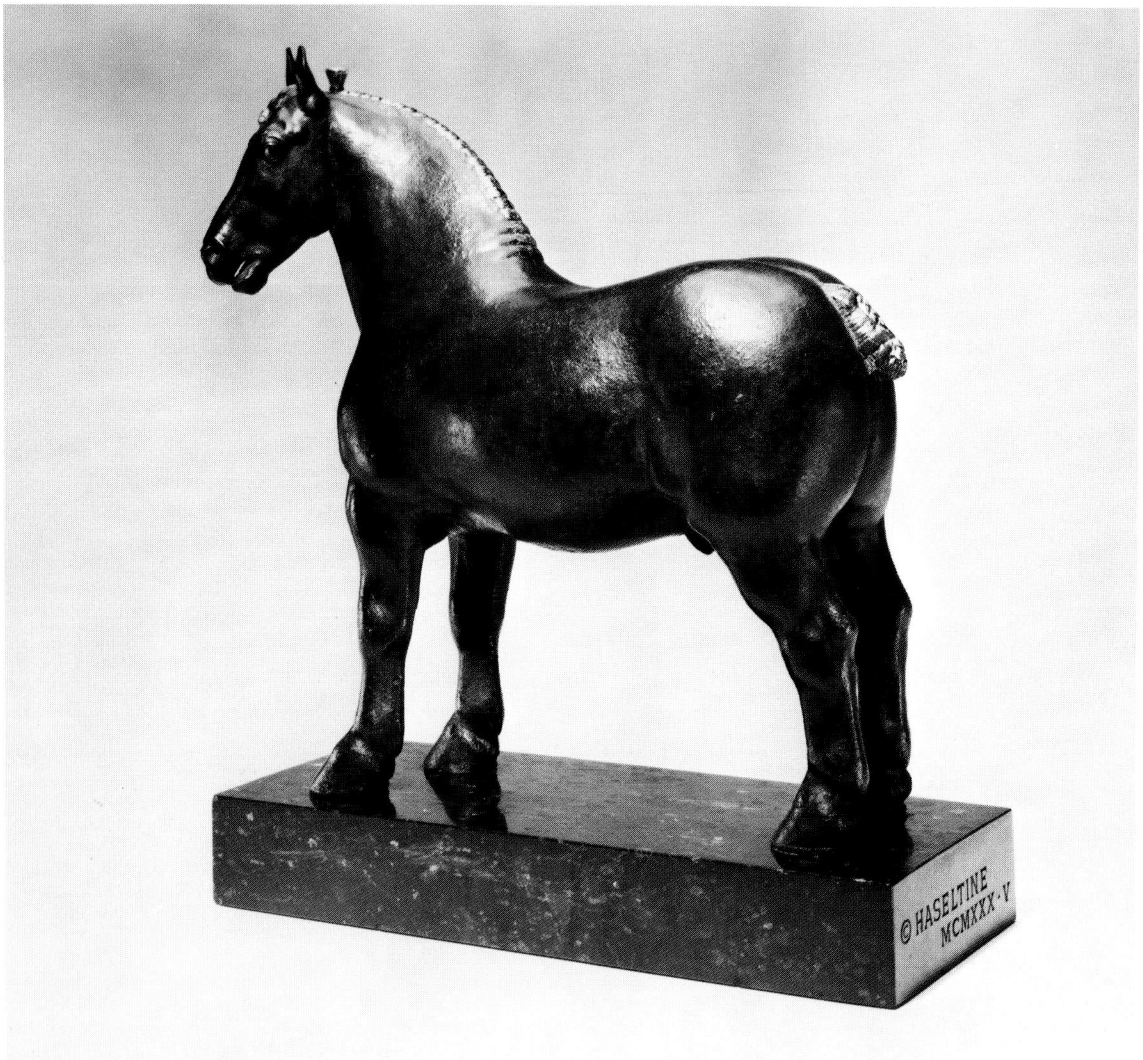

40.
Percheron Stallion: Rhum
1930

Cast bronze with lazuli, onyx, ivory and gold foil on Bardiglio marble base, 14⅝″ x 12¼″ x 5″ (including base)

Signed: © *Haseltine* (on base at right)

Dated and numbered: *MCMXXX•5* (on base at right)

Inscribed: *Percheron* (on base at left)

Lent by Conner-Rosenkranz, New York

GERTRUDE VANDERBILT WHITNEY

Born 1877, New York City
Died 1942, New York City

Gertrude Vanderbilt was the daughter of Cornelius Vanderbilt and the founder of the Whitney Museum of American Art. In 1896 she married Harry Payne Whitney and a few years later began her career as a sculptor. She studied with Hendrick C. Andersen and James Earle Fraser and at the Art Students League in New York. In Paris Mrs. Whitney was a pupil of Andrew O'Connor and Rodin, whose influence can be seen in the emotional expressionism and flickering surfaces of her bronzes. After returning to New York in 1907, she established a studio in Greenwich Village which became both a work space and meeting ground for the progressive American artists whose work she admired. The artist is best known for the several public monuments and fountains she designed, including the *Titanic Memorial* (1914) in Washington, D.C. and the equestrian *Buffalo Bill* (1924) in Cody, Wyoming.

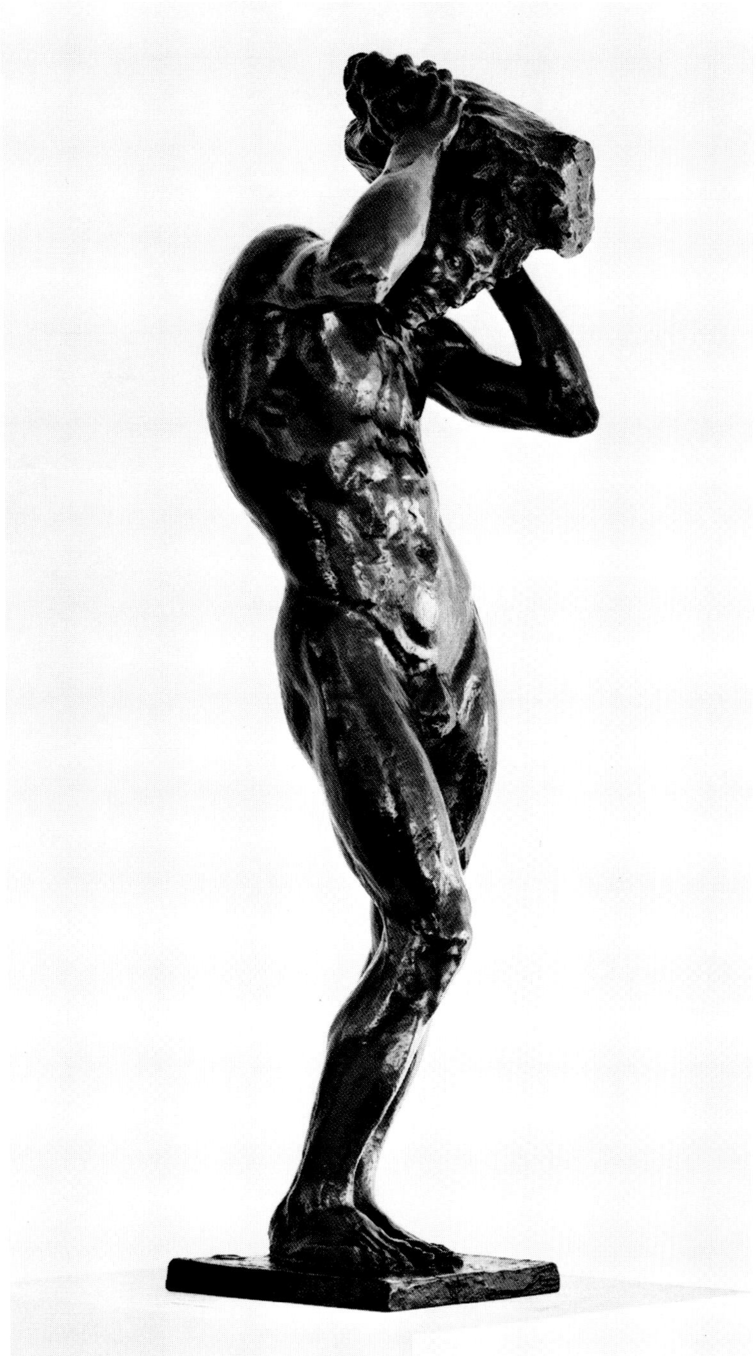

41.
A Caryatid
1913

Cast bronze, dark brown patina, H. 19″

Signed and dated: *Gertrude V. Whitney/1913* (on base at right)

Foundry mark: *C. Valsuani Cire Perdue* (stamped on base at back)

Lent by Robert Pincus-Witten and Leon Hecht, New York

This is a study for one of the three caryatids supporting a basin of fish heads and sea grapes forming a fountain originally designed for the Arlington Hotel in Washington, D.C. When shown at the Salon of 1913, the work won an honorable mention. The marble version of this fountain is now located at McGill University in Montreal.

MAHONRI MACKINTOSH YOUNG
Born 1877, Salt Lake City, Utah
Died 1957, Norwalk, Conn.

The grandson of Brigham Young, the artist was encouraged by his family to pursue his interest in drawing and modeling. In 1899, Young traveled to New York to study at the Art Students League. Two years later he continued his work in Paris at the Académie Julian and Académie Delecluse, but it was not until a short trip to Italy that he began to work primarily as a sculptor. After injuring his thumb while boxing, Young turned briefly to watercolor. In Paris the artist modeled the first of his figures of male workers that remain his finest contribution of American sculpture. Like the paintings of his contemporaries of the "Ashcan School," Young's sculptures use the laborer as subject matter but contain no overt social message. They are far removed from the politically sensitive images of downtrodden peasant laborers created by the French sculptors Dalou or Meunier. After his return to the United States, the artist incorporated other subjects into his work, including American Indians (inspired by a 1912 trip to Arizona), animals and boxing themes, as well as portraits. Young designed several public monuments and wrote frequently on art.

ABASTENIA St. LEGER EBERLE
Born 1878, Webster City, Iowa
Died 1942, New York City

Another artist best known for urban and working class subjects is Abastenia Eberle. After arriving in New York in 1899, she attended classes at the Art Students League with the sculptor George Grey Barnard and frequented the studio of Gutzon Borglum. For a time Eberle shared a studio with Anna Hyatt Huntington. They occasionally collaborated on sculptures, such as the *Men and Bull* shown at the St. Louis exposition (1904), for which the latter modeled the animals. In 1907 and 1908 Eberle worked in Italy, and in 1913 was in Paris. Earlier, around 1905, she had aligned herself with the progressive, non-academic artists of the period by depicting subjects, mostly women and children, drawn from the poor working class of New York City's Lower East Side. In 1914, she established her studio in this neighborhood. Although considered radical because of their unidealized subject matter, Eberle's figures often contain a tenderness and sentimentality that is more obvious in the work of her contemporary, Bessie Vonnoh. After 1919, the artist was in bad health and confined her work to occasional small bronzes and garden figures.

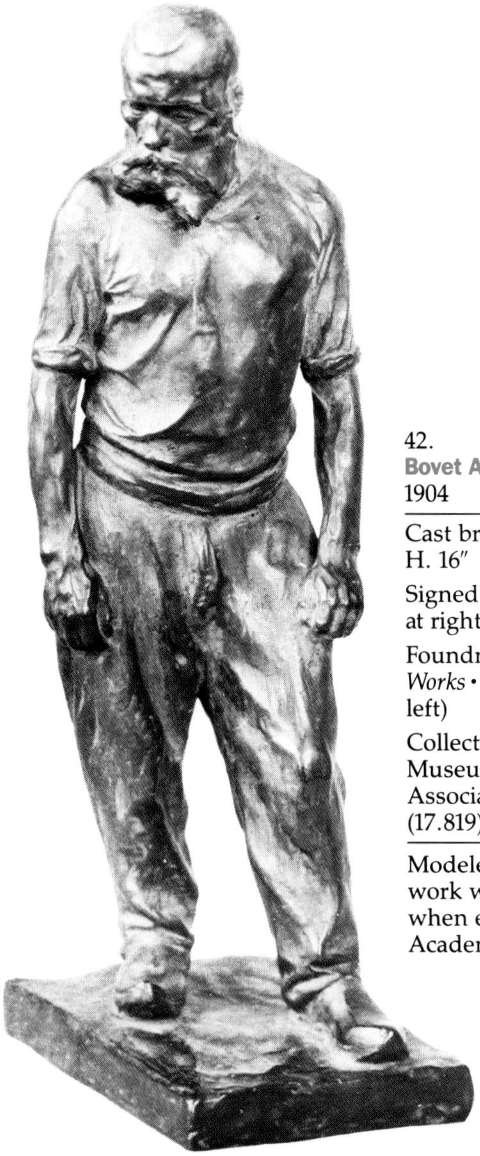

42.
Bovet Arthur–A Laborer
1904

Cast bronze, dark brown patina, H. 16″

Signed: *M.M. Young* (on back at right)

Foundry mark: *Roman • Bronze • Works • N.Y.* (on back at bottom left)

Collection of The Newark Museum. Purchase 1917 Associated Membership Dues (17.819)

Modeled in Paris in 1904, this work won the Barnett Prize when exhibited at the National Academy of Design in 1911.

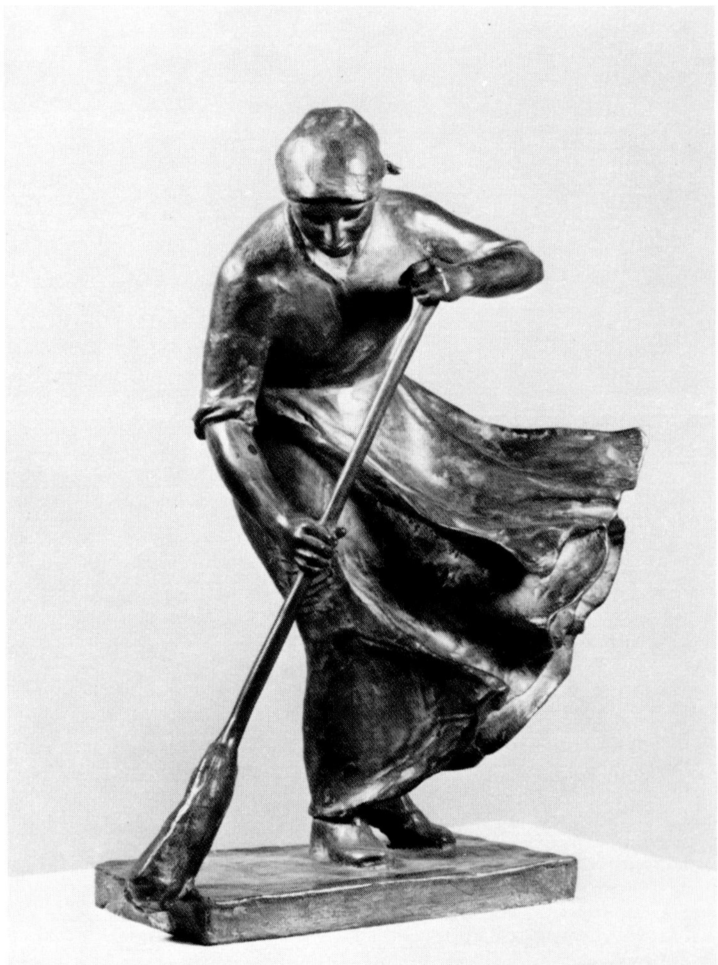

CHARLES CARY RUMSEY

Born 1879, Buffalo, N.Y.
Died 1922, Long Island, N.Y.

Rumsey came from a wealthy and socially prominent family. In 1893 he traveled with his family to Europe, where he decided to remain in Paris as a student assistant to the American sculptor Paul Weyland Bartlett. After returning home in 1895, he completed his formal education and continued to study sculpture at the Boston School of Fine Arts with Bela Pratt. Rumsey returned to Paris in 1902 and was greatly influenced by his studies with Frémiet, a specialist in equestrian statues who placed special emphasis on animal anatomy. The artist remained in Paris four years before returning to the United States and establishing a studio in New York City. He was an avid polo player and sportsman and many of his sculptures are related to these interests. Rumsey's first important public sculpture was the equestrian *Pizarro* modeled for the Panama-Pacific Exposition in San Francisco (1915), for which he won a bronze medal. The artist's career was briefly interrupted when he served in the cavalry in World War I. During the last years of his life his sculptural style became more monumental and abstract, reflecting the influence of the French sculptor Maillol. Rumsey died suddenly in an automobile accident.

43.
The Windy Doorstep
1910

Cast bronze, dark green patina, H. 14″

Signed and dated: *A St L. Eberle/1910* (on back left corner)

Foundry mark: *S. Klaber & Co./ Founders, N.Y.* (stamped on back at left)

Collection of The Newark Museum. Gift of Joseph S. Isidor 1913 (13.304)

This sculpture won the Barnett Prize when exhibited at the National Academy of Design in 1910.

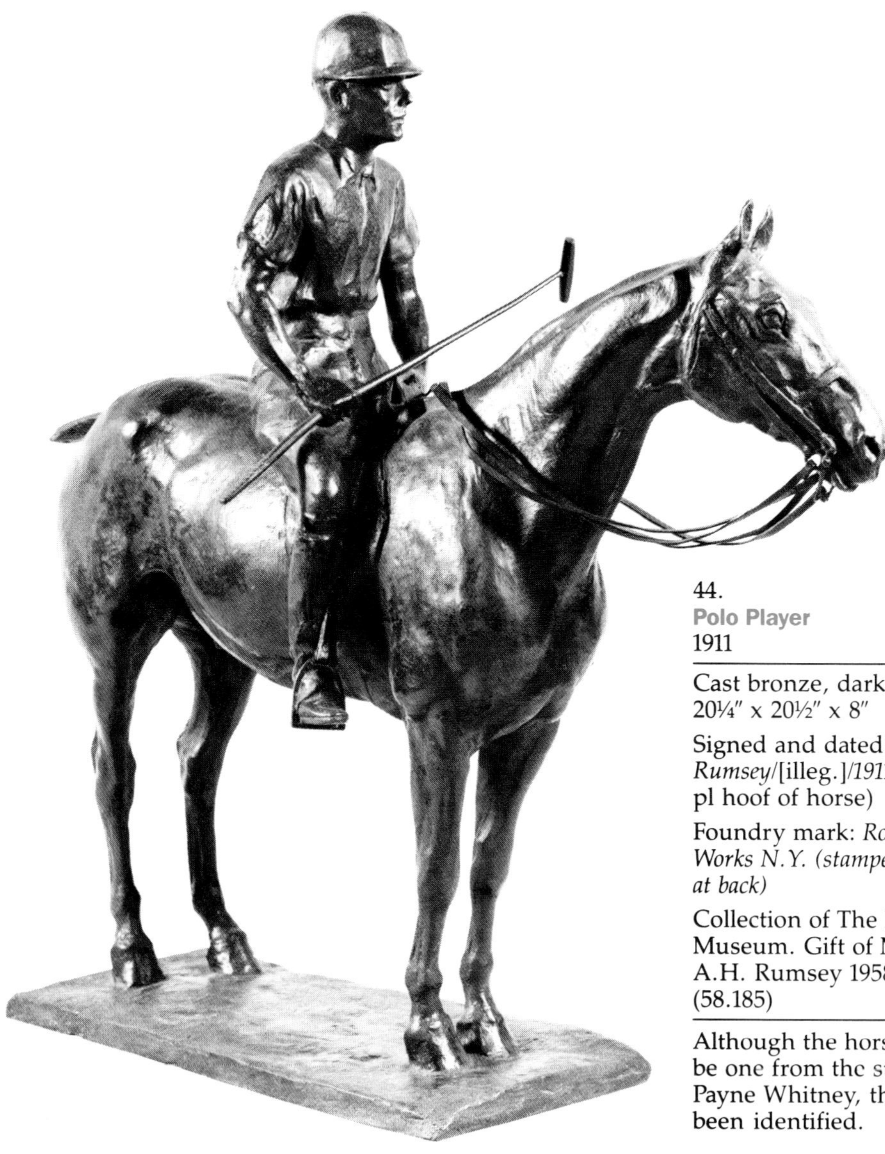

44.
Polo Player
1911

Cast bronze, dark brown patina, 20¼″ x 20½″ x 8″

Signed and dated: *C.C. Rumsey*/[illeg.]/*1911* (on base near pl hoof of horse)

Foundry mark: *Roman Bronze Works N.Y. (stamped on base at back)*

Collection of The Newark Museum. Gift of Miss Mary A.H. Rumsey 1958 (58.185)

Although the horse is thought to be one from the stables of Harry Payne Whitney, the rider has not been identified.

HARRIET WHITNEY FRISHMUTH
Born 1880, Philadelphia, Pa.
Died 1980, Southbury, Conn.

Frishmuth spent much of her early life traveling in Europe. At the age of nineteen, she entered Rodin's sculpture class for women and learned his fluid, suggestive modeling of the figure. She also studied with Injalbert and Henri-Desire Gauquie and spent two years working in Berlin with Cuno von Euchtritz. After returning home, Frishmuth continued her studies at the Art Students League with Borglum and MacNeil. She worked for a short period as an assistant to Borglum and Bitter before establishing her own studio in New York. Frishmuth was well known for her fountain sculptures, usually depicting nude female figures in spirited poses. Although she always retained the sensuous modeling learned from Rodin, by the 1920s Frishmuth's work exhibited a crisp simplification of form that is closely related to the decorative arts of that period. Enjoying a long and successful career, the artist won numerous awards and medals.

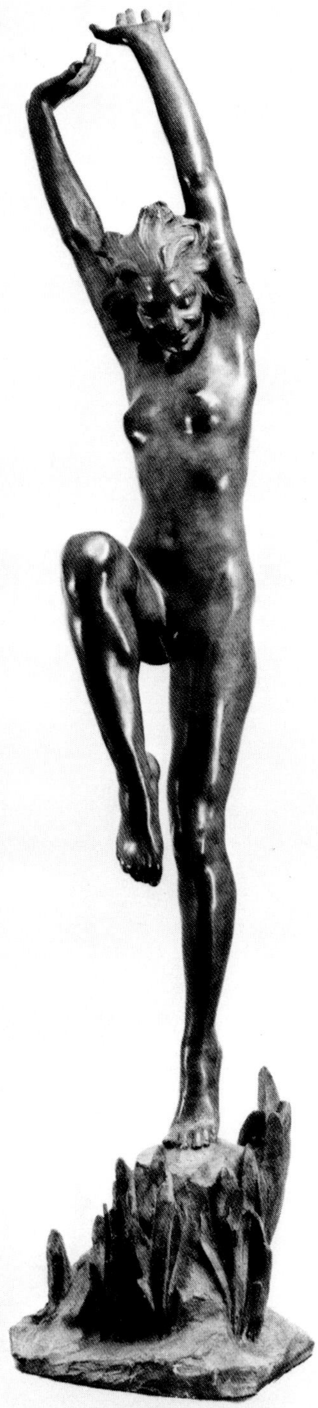

45.
Joy of the Waters
1920

Cast bronze, green patina, H. 44″

Signed and dated: *Harriet W. Frishmuth Sc/©1920* (on back of base at right)

Foundry mark: *Gorham Co. Founders/QBKX* (stamped on base at back)

Lent by the National Academy of Design, New York.

CHESTER BEACH

Born 1881, San Francisco, Calif.
Died 1956, Brewster, N.Y.

After graduating from Lick Polytechnic School, Beach worked as a silver designer and attended night classes at the Mark Hopkins Institute. He lived briefly in New York before going to Paris in 1904. There Beach studied at the Académie Julian and maintained his own studio. Like many of his contemporaries, Beach was greatly influenced by the work of Rodin. After returning to New York in 1907, the artist concentrated on a series of bronze statues of laborers which are related to the work of the French sculptor Jules Dalou and the urban realism of the American painters of the "Ashcan School." After spending 1911 and 1912 in Rome, Beach's style became more classical and his subject matter more allegorical. Many of his statues are related to the movement of water, either in their subjects (i.e. *The Surf*, 1921; *Rising Sea Mists*, 1925) or their use as fountains. The artist worked frequently as a medalist and in a variety of materials, including marble and ivory. Beach won numerous awards and was a president of the National Sculpture Society.

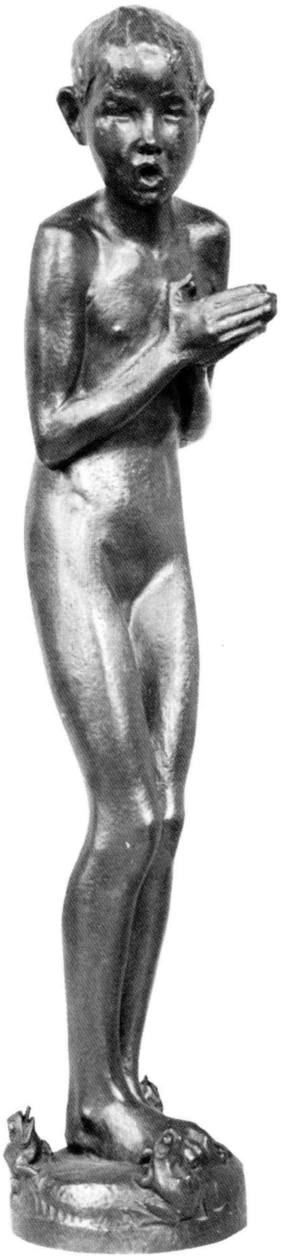

46.
Swimmin'
1907

Cast bronze, brown patina, H. 38¼"

Signed: *C. Beach*/© (on base near pr foot)

Foundry mark: *Kunst-Foundry-N.Y.* (on back of base)

Collection of The Newark Museum. Purchase 1927
J. Ackerman Coles Bequest Fund
(27.288)

This piece first appeared as a small bronze and later in this fountain form, of which three were cast.

ALICE MORGAN WRIGHT

Born 1881, Albany, N.Y.
Died 1975

After graduating from Smith College in 1904, Wright studied at the Art Students League with Gutzon Borglum, MacNeil and James Earle Fraser. Five years later, after winning the League's Borglum and Saint-Gaudens prizes, she left for Paris and further study at the Ecole des Beaux-Arts and Académie Colarossi. While in Europe the artist became involved with the Women's Suffrage movement, one of several social and humanitarian causes she championed throughout her life. Because of these activities and her thoughtful approach to each sculpture, Wright was not a prolific artist and few of her sculptures are known. She was one of the first American sculptors to explore the possibilities of Cubist and non-objective forms, although her advanced and traditional work existed side by side.

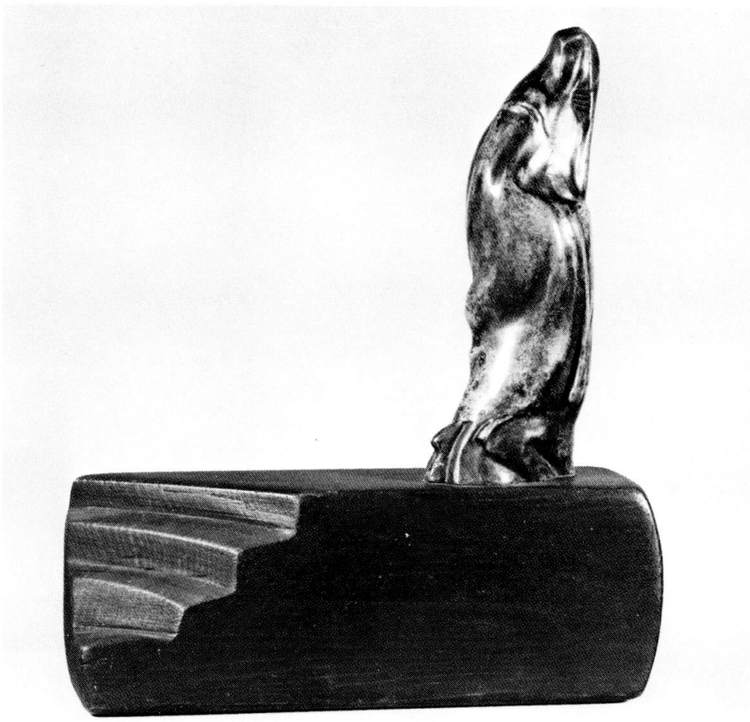

47.
Lady MacBeth
1920

Cast bronze, dark green patina; wood base, 12⅜" x 11" x 2⅞"

No markings (cast by Kunst Art Foundry, N.Y.)

Collection of The Newark Museum. Gift of Joseph S. Isidor 1923
(23.1860)

CECIL De BLAQUIERE HOWARD
Born 1888, Clifton, Canada
Died 1956, New York City

The son of accomplished amateurs in the arts, Howard was raised in Buffalo, New York, and received his first training in sculpture from James Earle Fraser, who briefly taught in that city. After his father's death, Howard moved with his mother to Paris in 1905. He studied at the Académie Julian with Raoul Verlet and, from 1909 to 1910, in Antwerp with the popular *animalier* Rembrandt Bugatti. The group of animal studies he exhibited at the Salon d'Automne that year won him membership in the Société des Artistes Francais. Until 1940, when he was forced by the advancing German army to flee to the United States, Howard lived in Paris and enjoyed a successful career as an expatriate sculptor. His small bronzes, of which boxing groups form a large portion, were usually modeled in wax rather than clay, and cast directly in editions of four. This process eliminated the need for a plaster model and retained more of the spontaneity of the artist's touch. Howard first came to the attention of the American public when he exhibited in New York City in 1915 and 1916. He enjoyed a successful career in both Europe and America and in his later works often showed a preference for direct carving in stone.

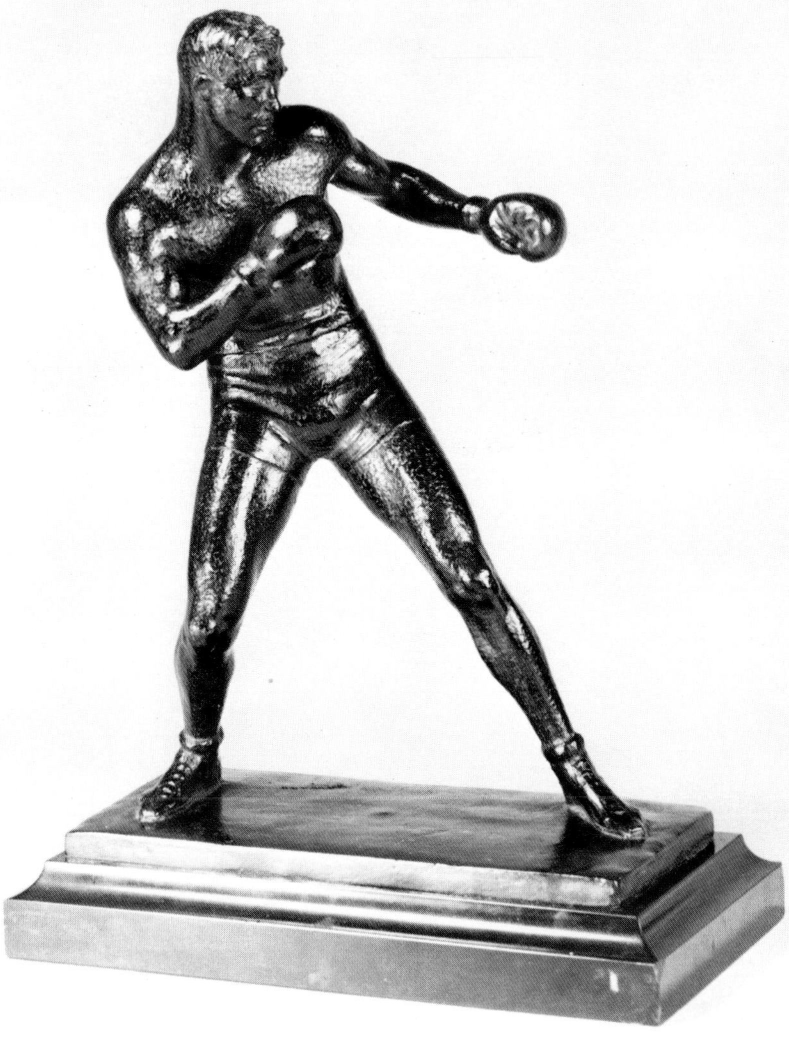

48.
Boxer–American Style
ca. 1935

Cast bronze, dark brown patina, H. 12⅛″; black stone base, 78″ x 9¾″ x 5½″

Signed and numbered: *Howard/1* (on base at left)

Foundry mark: *Bronze/Cire/ J. Valsuani/Perdue* (stamped on back of base)

Collection of The Newark Museum. Gift of Henry Luce III 1959
(59.415)

GASTON LACHAISE
Born 1882, Paris, France
Died 1935, New York City

The son of a prominent French woodcarver and cabinet maker, Lachaise studied at the Ecole Bernard Palissy with Jean-Paul Aube and Alphonse Moncel and at the Académie Nationale des Beaux-Arts with Gabriel Jules Thomas. About 1902, Lachaise met an American woman, Mrs. Isabel Dutaud Nagel, who became his mistress, model and, later, wife. After working for a brief period for the designer René Lalique, Lachaise followed Mrs. Nagel to Boston in 1906. For almost six years he worked as an assistant to the Boston sculptor, Henry Hudson Kitson, completing decorative details on large outdoor sculptures. Lachaise moved to New York City in 1912 and the following year became the chief assistant to Paul Manship, with whom he remained until 1921. While earning his living as an assistant to others, Lachaise continued to develop his own sculptural style, with its increasing interest in mass and volume. Working most often with the female form, he condensed the human body to its essential geometric forms, emphasizing flowing lines and voluptuous curves. The artist's first public, if controversial, recognition came when the Stephan Bourgeois Gallery in New York gave him a solo exhibition in 1918. Between 1931 and 1935, Lachaise received several large-scale commissions, including reliefs for the RCA and International buildings in Rockefeller Center. During this period, his non-public sculptures became more explicitly sexual and anatomically distorted. Lachaise was honored with a retrospective at the Museum of Modern Art, its first of a living American artist. He died of leukemia only a few months later.

49.
Peacocks
1922

Cast bronze, dark brown patina with traces of gilding, 22¼″ x 55¾″ x 9″ (including integral black stone base)

Signed and dated: *G Lachaise/©1922* (on top of base at right)

Foundry mark: *Roman Bronze Works N-Y-* (on top of base at back)

Collection of The Newark Museum. Gift of Mr. and Mrs. Franklin Conklin, Jr. 1952 (52.96)

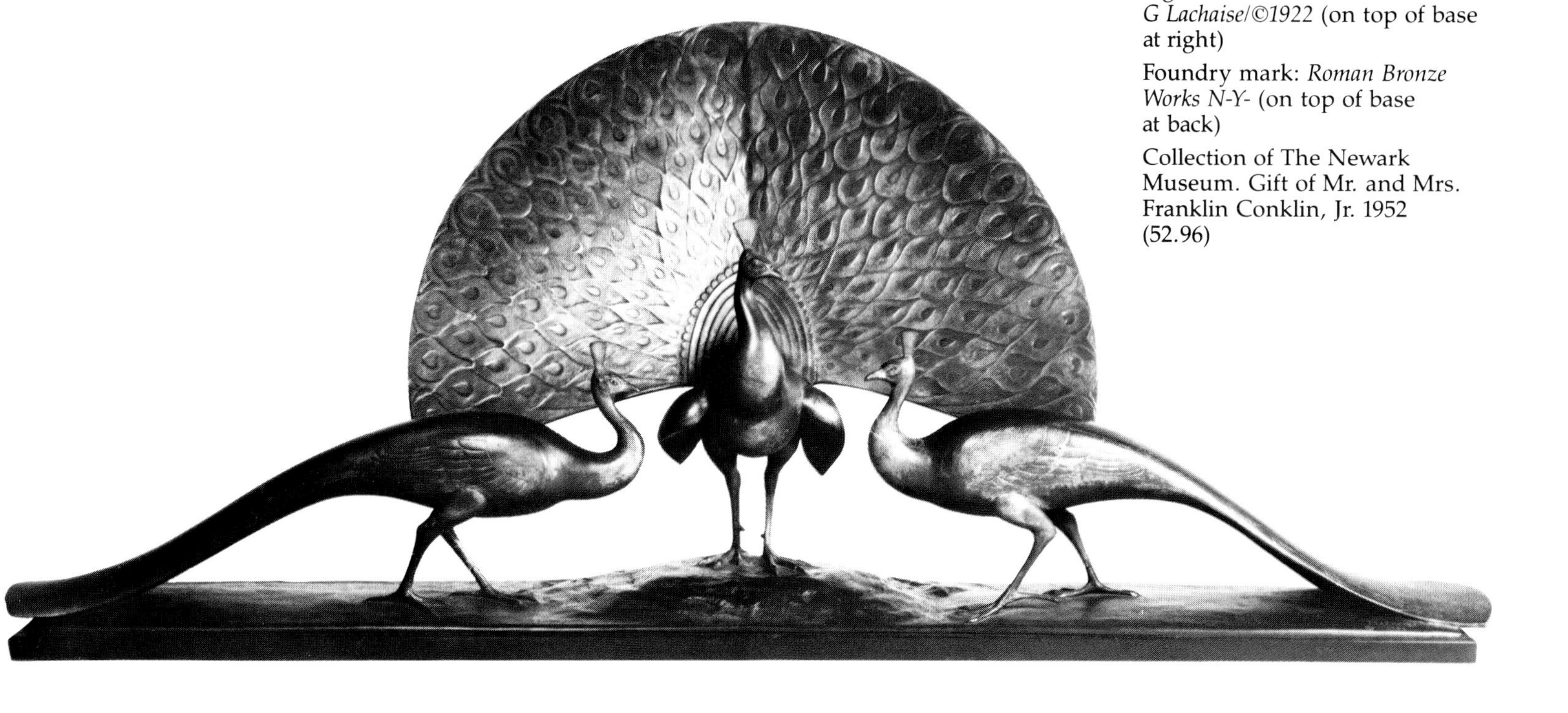

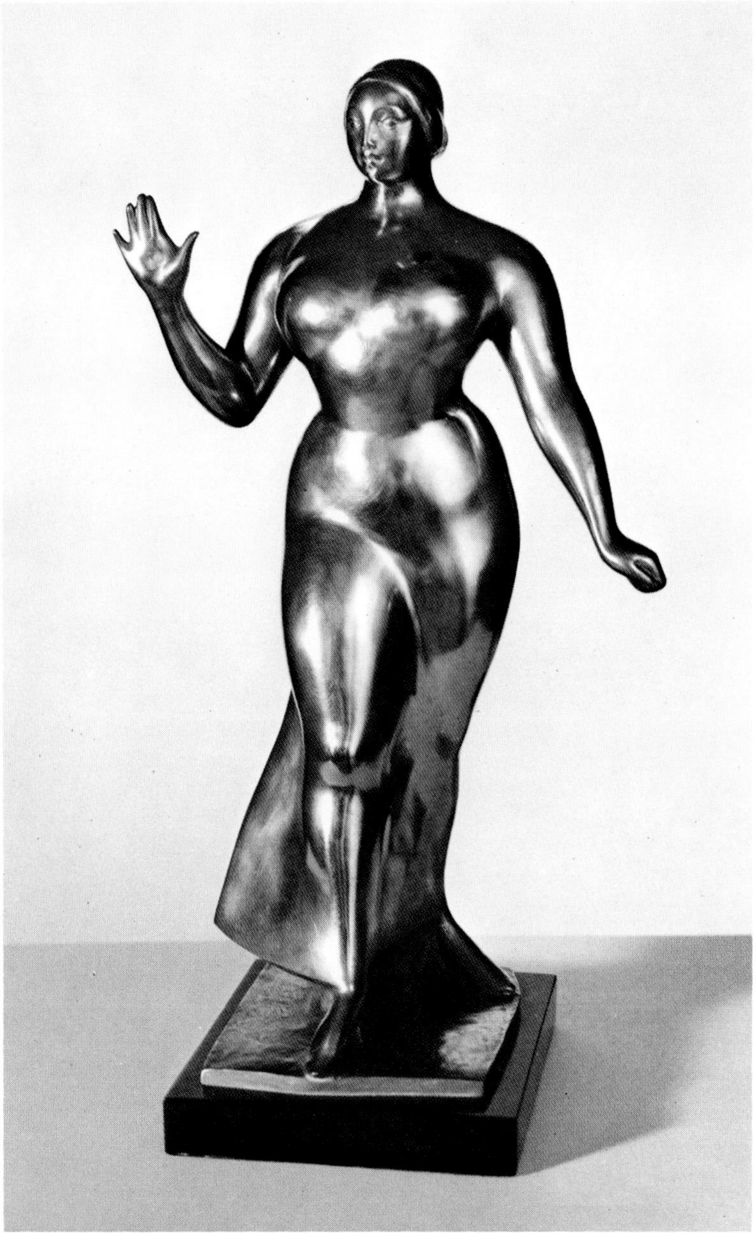

50.
Walking Woman
1922

Cast bronze, polished, 18¼″ x 10½″ x 6½″; black stone base, 1⅛″ x 6¼″ x 7¼″

Signed and dated:
G. Lachaise/©1922 (on top of base at back)

No foundry mark

Collection of The Newark Museum. Purchase 1938 Exhibit Purchase Fund
(38.571)

51.
Head of a Woman
(Egyptian Head)
1923

Cast bronze, polished and lacquered, H. 13″ on black stone base 4¼″ x 8″ x 8″

Signed and dated:
G. Lachaise/© 1923 (on base underneath pl neck)

Foundry mark: *Rom*[an Bronze Works; illeg.] (on back)

Collection of The Newark Museum. Gift of Mr. and Mrs. Felix Fuld 1927
(27.1153)

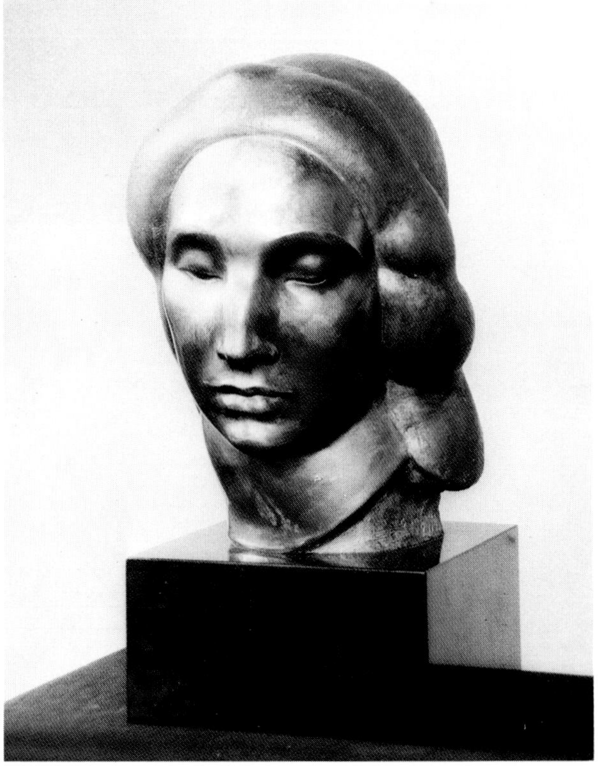

ELIE NADELMAN
Born 1882, Warsaw, Poland
Died 1946, Riverdale, N.Y.

By the time he came to the United States in 1917, Nadelman was already a well-known sculptor. After studying in Warsaw, Krakow and Munich, Nadelman arrived in Paris about 1903 and soon became a member of the group of avant-garde artists who centered around the drawing rooms of the American collectors Leo and Gertrude Stein. Although the vaguely modeled forms of his early work owed something to the influence of Rodin, Nadelman moved steadily toward a more abstract style of simplified geometric shapes and carefully analyzed volumes. For a short period, around 1907, his work paralleled and even anticipated the Cubist experiments of Picasso and Braque. For a 1910 issue of Alfred Steiglitz's magazine, *Camera Work,* Nadelman wrote, "The subject of any work of art is for me nothing but a pretext for creating significant form, relations of forms which create a new life that has nothing to do with life in nature, a life from which art is born, and from which spring style and unity." His reputation in the Parisian art world helped to make Nadelman a particularly influential sculptor in New York. His inventive approach to materials often led him to paint or inlay his sculptures, which might assume their forms in bronze, wood, stone, papier-mâché or terra-cotta. In 1919, Nadelman married a wealthy widow, Mrs. Joseph A. Flannery, and the couple bought an estate at Riverdale-on-Hudson. Before their fortunes were reversed in the stock market crash of 1929, they amassed a large collection of American folk art that filled their home. Finally, the sculptor lost his large studio and suffered the destruction of most of its contents. Although he continued to work, the last decade of Nadelman's life was spent in seclusion and almost complete isolation from the art world.

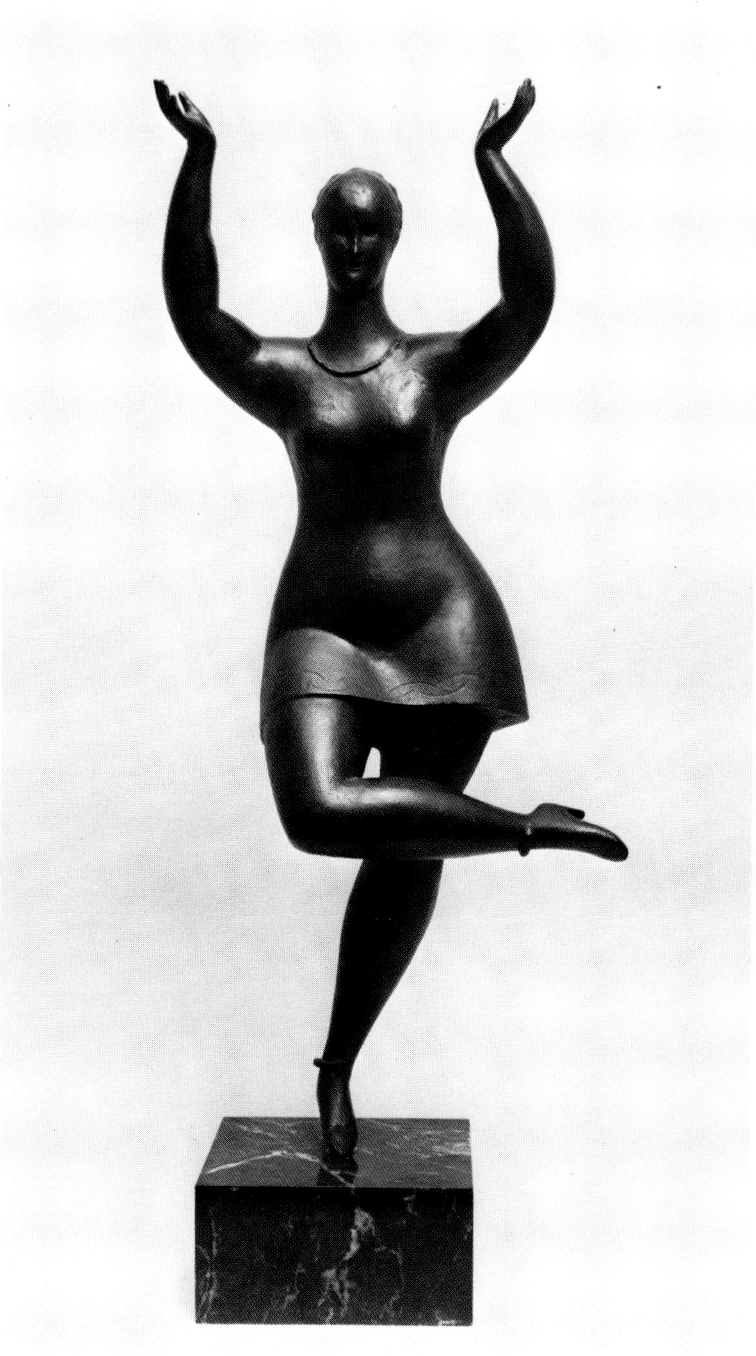

52.
Female Dancer
ca. 1920

Cast bronze, H. 32½″

Lent by Mr. and Mrs. E. Jan Nadelman

JO DAVIDSON
Born 1883, New York City
Died 1952, near Tours, France

In the years immediately following World War I, Davidson established his career as America's foremost portrait sculptor. The artist grew up in a poor family on the Lower East Side and, determined to be an artist, attended the Art Students League on scholarship. At his parents' urging, Davidson went to New Haven to seek admission to Yale University Medical School but ended up being given free instruction at the Yale Art School. Shortly thereafter he returned to New York and persuaded Hermon MacNeil to take him as an assistant. He remained with MacNeil until 1904 and in 1907 went to Paris where he ended his formal training with a brief period of instruction at the Ecole des Beaux-Arts. Before returning to New York in 1910, Davidson met Gertrude Vanderbilt Whitney, who became his great friend and patron. At the close of World War I, the artist modeled a series of portrait busts of the allied leaders. In the following decades his sitters included a virtual "Who's Who" of world politicians, writers, artists, businessmen and military leaders. Much of his career was spent working in France, where he was made a *chevalier* of the Legion of Honor.

WILLIAM HUNT DIEDERICH
Born 1884, Szent-Grot, Hungary
Died 1953, Tappan, N.Y.

Diederich's mother was a daughter of the Boston painter William Morris Hunt and after his father's death she took him to live in the United States. He attended school in Boston and then spent two years working on ranches in the West. After his return East, Diederich entered the Pennsylvania Academy of the Fine Arts, where he met Paul Manship, a fellow student in Charles Grafly's sculpture class. In 1907, he and Manship took a walking tour of Spain. The following year he won the Academy's Stewardson Prize and left for Europe. Diederich traveled widely in Africa and Europe, spending ten years working in Paris. In addition to his bronze sculpture, the artist was known for his work in wrought iron. A skilled craftsman and designer, he produced a host of decorative objects, including weathervanes, fire screens and chandeliers (an example of which is owned by The Newark Museum). Diederich also received several commissions for architectural decorations, including a *Pegasus with Messenger* for the post office in Westwood, New Jersey.

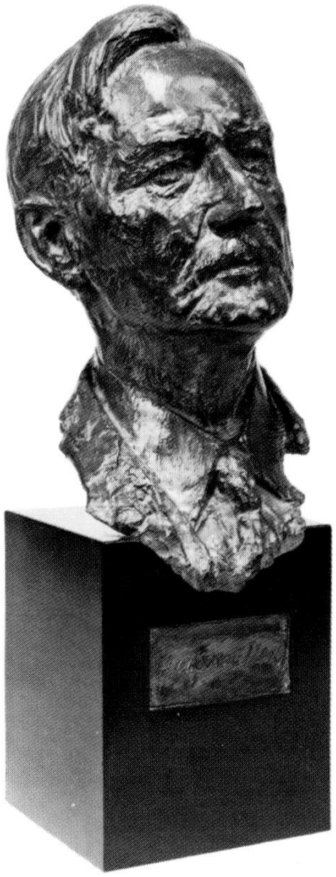

53.
Portrait of Hermon Atkins MacNeil
1945

Cast bronze, brown patina, on Macarta base, 22¼″ x 6¾″ x 8″ (including base)

Signed and dated: *Jo Davidson 1945* (on back of neck)

Bronze plaque bearing sitter's signature inset on front of base.

No foundry mark

Lent by Conner-Rosenkranz, New York

This is a unique cast of the portrait bust modeled two years before MacNeil's death.

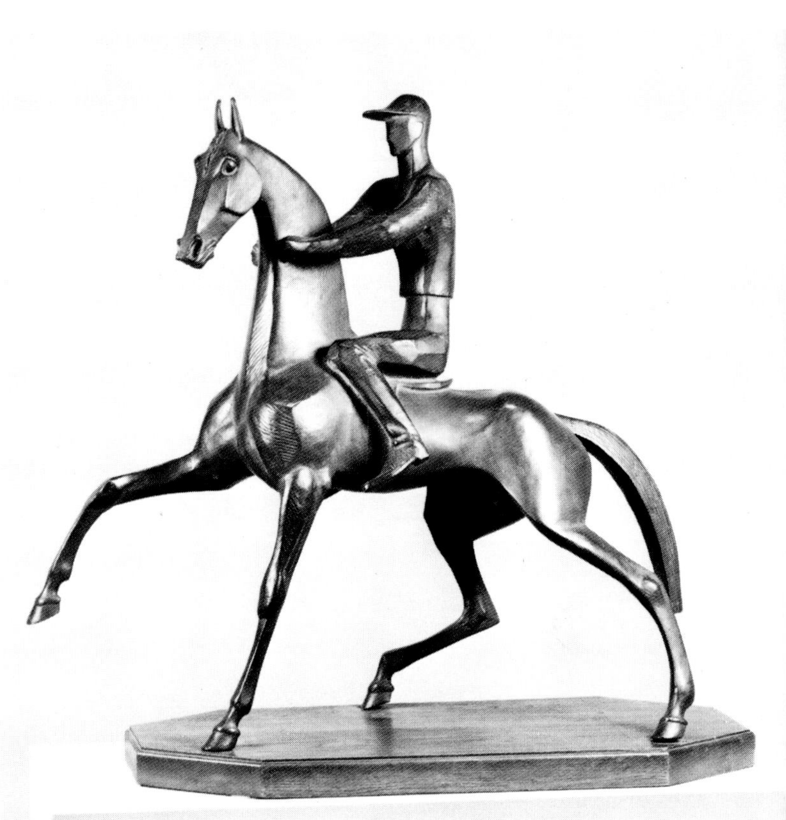

54.
The Jockey
1924

Cast bronze, dark green patina, 23¼″ x 24¼″ x 10½″; wood base, 1½″ x 22½″ x 11″

Signed and dated: *H Diederich 1924* (underneath horse's tail)

No foundry mark

Collection of The Newark Museum. Gift of Mr. and Mrs. Felix Fuld 1927 (27.1073)

PAUL HOWARD MANSHIP

Born 1885, St. Paul, Minn.
Died 1966, New York City

After early training at the St. Paul Institute of Art and work as an engraver, Manship moved to New York City in 1905. He studied briefly at the Art Students League with Jo Davidson and then worked as an assistant in Solon Borglum's studio. In 1906 and 1907 he studied under Charles Grafly at the Pennsylvania Academy, an experience he later felt important to his development as a sculptor. The following year, the artist returned to New York to work in the studio of Isidore Konti. His earliest recognition came in 1909, when he won the Prix de Rome and spent three years at the American Academy in Rome. There, he was inspired by ancient Greek and Roman sculpture, especially of the Archaic period. When he returned to the United States in 1913, his style was characterized by a rigid frontality and simplicity of form that reflected this influence. Manship's brand of archaic classicism, with its strong decorative appeal, was instantly popular in this country and made him one of the most honored sculptors of his day. He won almost every award his profession offered and served as president of the National Sculpture Society. For many critics, Manship's work represented the perfect marriage of the academic tradition with "modern" abstract tendencies. During the 1920s, the artist worked primarily in his studio in Paris and taught at the American Academy in Rome. Many of Manship's important commissions were for garden or architectural sculptures, including the *Prometheus* fountain for Rockefeller Center.

55.
The Moods of Time: Day
1938

Cast bronze, green patina, 48″ x 85½″; mounted on stone base

Signed and dated: *-Day-/Paul Manship/Sculptor/1938* (on front on cloud)

Lent by the Equitable Life Assurance Society

This is one of four fountain figures that Manship modeled for the New York World's Fair, 1939. He once wrote that the figure of *Day* symbolized "energy, radiation, speed."

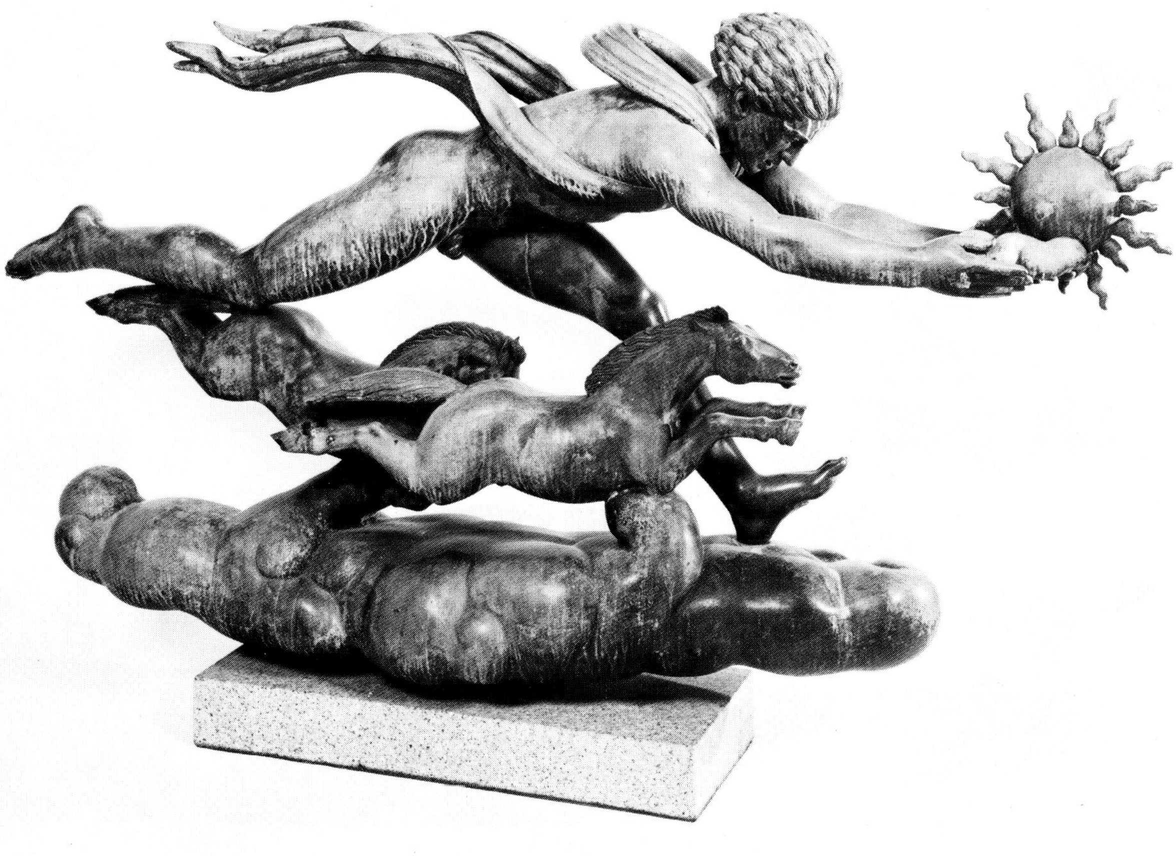

MALVINA CORNELL HOFFMAN
Born 1885, New York City
Died 1966, New York City

The daughter of the pianist Richard Hoffman, many of Malvina Hoffman's most successful sculptures were of musicians and dancers. Although she originally intended to be a painter, the artist received instruction from Herbert Adams, George Grey Barnard and Gutzon Borglum before leaving for Paris in 1910. After persuading Rodin to take her as a student, Hoffman returned to New York at his advice to study anatomy for a year at the Cornell College of Physicians and Surgeons. She then worked in Paris until the outbreak of World War I. While on a short visit to London in 1911, Hoffman saw a performance by the Russian dancers Anna Pavlova and Mikhail Mordkin that inspired her to begin a series of dance subjects. Hoffman became a particularly close friend of Pavlova and in the following years made numerous sculptures of her dancing. The beautifully undulating surfaces of her early works show the profound influence of Rodin. In 1924, Hoffman married the English violinist Samuel Grimson, and three years later traveled to Yugoslavia to study sculpture with Ivan Mestrovic. As Hoffman's career developed, her style became less indebted to Rodin and took on a simplified naturalism. She worked frequently as a portraitist and her largest project was the series of 105 figures representing the races of the world begun in 1930 for the Field Museum of Natural History in Chicago. The artist was active in relief work during both world wars and was selected as the sculptor for the American War Memorial Building in Epinal, France.

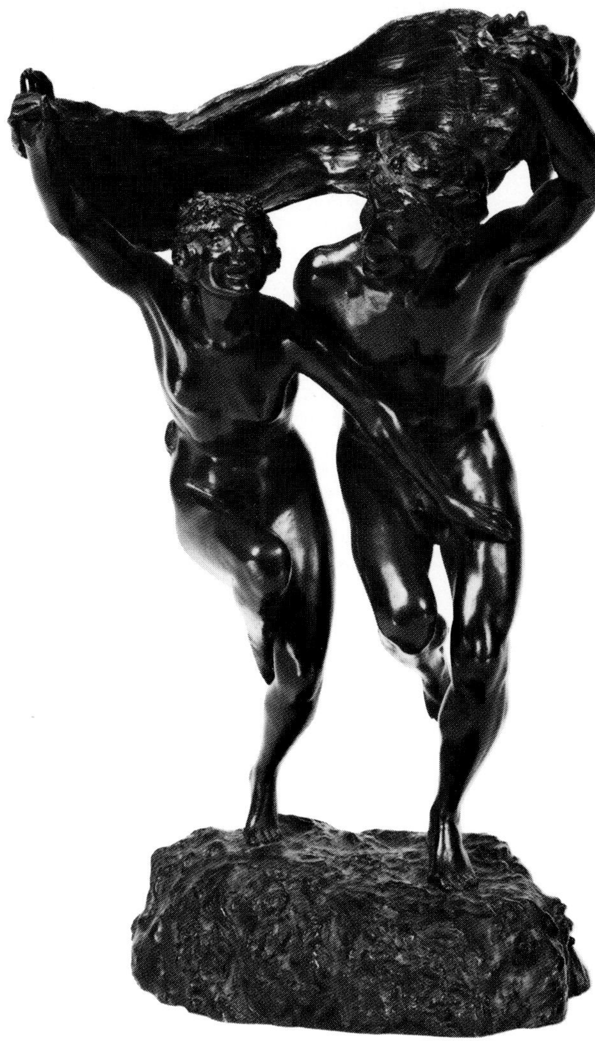

57.
Nijinsky in "L'Après-midi d'un faune"
1912

Cast bronze, dark brown patina, 8⅝" x 16½" x 4½"

Signed: *Malvina Hoffman* © (on back)

Foundry mark: *Roman Bronze Works Inc. NY* (on base at left)

Lent by the Malvina Hoffman Estate, courtesy of Berry-Hill Galleries, New York

Although Nijinsky did not pose for her, Hoffman saw him dance in Paris in 1912. This is the second of two lifetime castings of the piece.

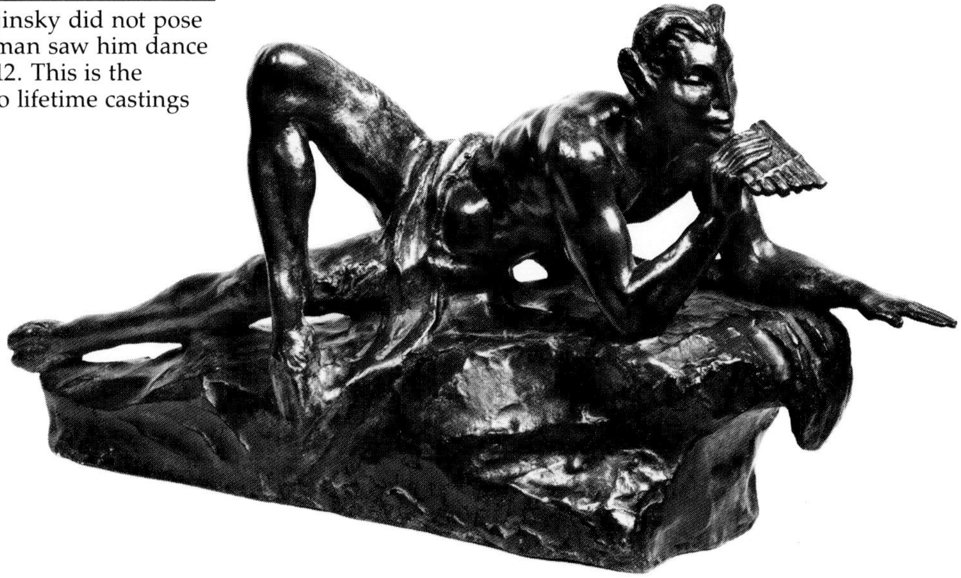

56.
Bacchanale Russe
1912

Cast bronze, dark brown patina, 14½" x 8" x 8½"

Signed: *Malvina Hoffman* © (on base at rear)

Foundry mark: *Roman Bronze Works N-Y-* (on base at rear)

Lent by Conner-Rosenkranz, New York

This sculpture represents Pavlova and Mordkin nude as they make their entrance to the "Autumn" section of Glazunov's *The Seasons* in the dance *Bacchanale.* It won a first prize when shown in the annual exhibition of the Société Nationale des Beaux-Arts in 1912.

LEO FRIEDLANDER
Born 1888, New York City
Died 1966, White Plains, N.Y.

From the age of fourteen to twenty, Friedlander worked for Klee Brothers, a firm that produced ornamental details in a variety of materials for buildings. Although he studied abroad at art schools in Brussels and Paris from 1908 to 1911, it was his early training that influenced the strongly architectonic quality of his mature work. In 1913, Friedlander won the Prix de Rome for sculpture and spent three years at the American Academy in Rome. Like Paul Manship, he was tremendously influenced by antique sculpture, with its emphasis on frontality and closed, self-contained forms. During the 1920s, Friedlander worked in New York as an assistant to Manship and Hermon A. MacNeil, while establishing his own career with several commissions for garden and decorative sculptures. In 1929, the artist received the important commission from the firm of McKim, Meade and White for a pair of monumental statues for the entrance of the Arlington Memorial Bridge in Washington, D.C. In the years preceeding World War II, Friedlander produced a host of public sculptures including his *Radio* and *Television* reliefs for Rockefeller Center and the *Four Freedoms* for the central mall of the 1939 New York World's Fair. Although he served as president of the National Sculpture Society from 1954 to 1957 and produced several large monuments, Friedlander's last years were spent in relative obscurity and at odds with contemporary abstract sculpture.

58.
Valor
1915

Cast bronze, dark brown patina, 30¼″ x 29″ x 13½″

Signed: *Friedlander* (on top of base at right)

Foundry mark: *Kunst Foundry N.Y.* (on top of base at left)

Lent by Conner-Rosenkranz, New York

Commissioned for the entrance to the Arlington Memorial Bridge in 1929, *Valor* (modeled 1915) and its companion group *Sacrifice* were caught in several governmental delays and not installed until 1951. At one point in the 1930s, the artist was asked to rework the groups so that they could be carved in granite. Finally, in 1949, the Italian government paid for the casting and gilding of the nineteen-foot bronze sculptures in appreciation for America's assistance in the Marshall Plan.

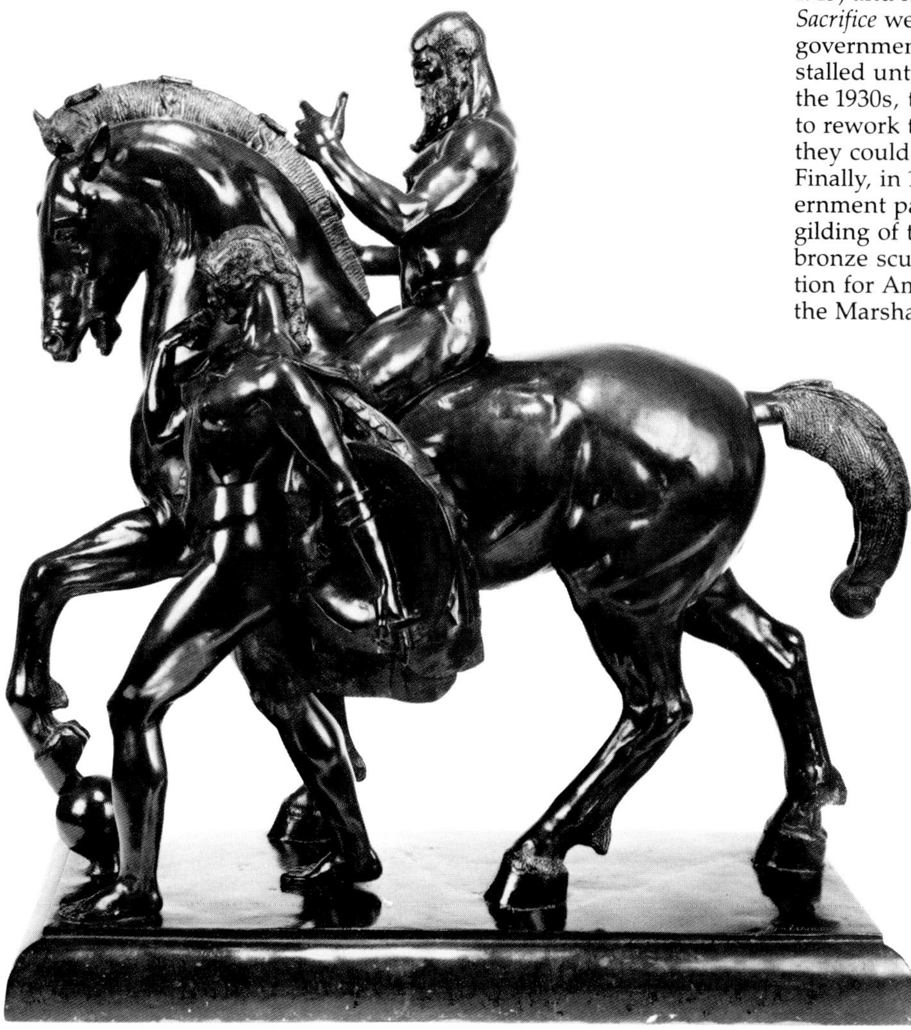

WILLIAM ZORACH

Born 1889, Eurburg, Lithuania
Died 1966, Bath, Maine

Zorach's first career was as a painter, for which he studied in both New York and Paris. After experimenting with wood and ceramic sculpture in the late 1910s, Zorach abandoned painting altogether in 1922 to begin his career as a sculptor. Zorach stressed the relationship between the artist and his material, becoming one of the pioneers in American of direct carving in stone and wood. The artist used his austerely stylized and monumental style to interpret individual human or animal subjects and the emotional interplay between man and woman. Although many of his earlier works were later cast in bronze, Zorach seems to have worked only occasionally in the material. By his death, he was considered one of the early leaders of modern sculpture in this country.

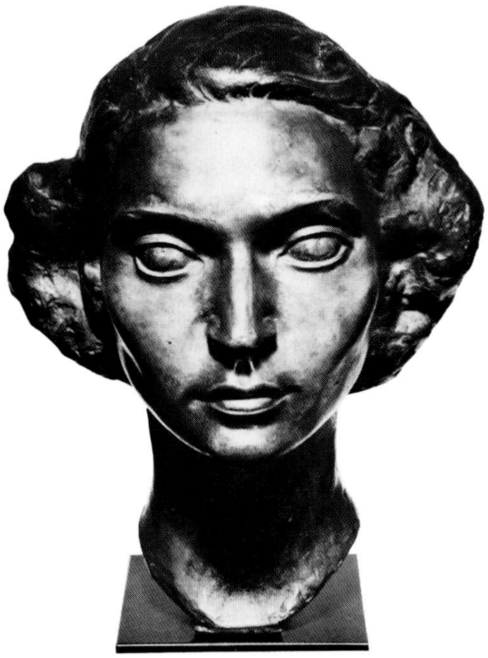

59.
Portrait of Rita K. Hillman
1935

Polished bronze, H. 14¾″; black marble base, 5″ x 6″ x 6″

Signed and dated: *Zorach/1935* (on back at center)

Anonymous loan

The sitter is the wife of the late publisher and art collector, Alex Hillman.

CARL PAUL JENNEWEIN

Born 1890, Stuttgart, Germany
Died 1978, New York City

Before immigrating to New York in 1907 in order to work for an architectural firm, Jennewein had apprenticed as a museum technician, learning painting, modeling and casting, and had also studied art history. He took classes at the Art Students League while working for a firm that supplied architectural sculpture and, in 1912, was awarded the Avery Prize by the New York Architectural League. This allowed Jennewein to travel and study for two years in France, Germany, Italy and Egypt. After returning to the United States, he won the Prix de Rome in 1916. Like that of Paul Manship and Leo Friedlander, Jennewein's style of decorative classicism was formed while in Rome as a student at the American Academy. Jennewein was particularly interested in the linear rhythm of Hellenistic ornamental sculpture and sought to give his own work a similar grace and charm. Following service in the American Red Cross during World War I, the artist established his studio in New York and quickly became one of the most sought-after sculptors for large-scale monumental and architectural work. In 1932, Jennewein and Leon V. Solon executed the colored terra-cotta sculptural decorations for the exterior of the Philadelphia Museum of Art. Jennewein led a busy and well-rewarded career and much of his work adorns government buildings in Washington, D.C.

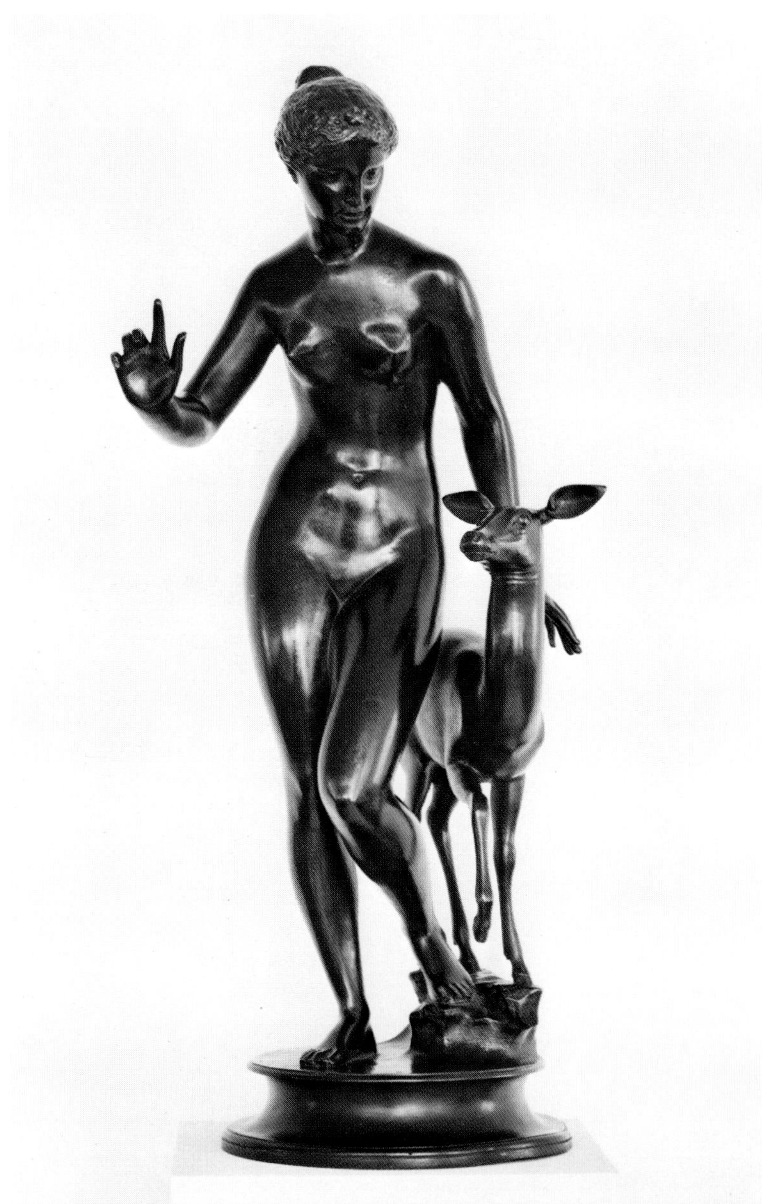

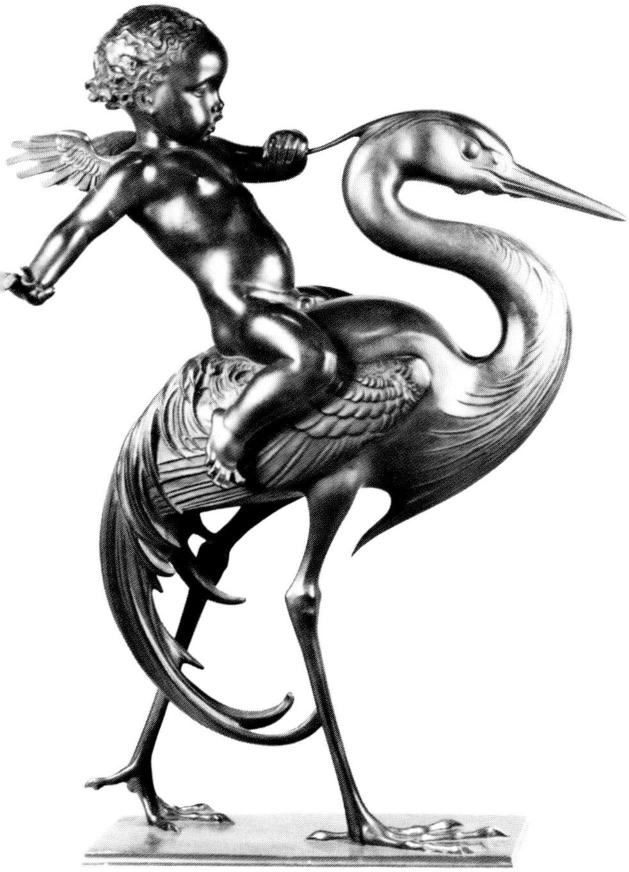

60.
Diana and Gazelle
1922

Cast bronze, dark brown patina, H. 23″

Signed and dated: • *C • P • Jennewein • S • C • © • 1922* (on front of base)

Foundry mark: *A Kunst Bronze Foundry, N.Y.* (on back)

Inscribed: *N° 3* (on back beneath rock)

Lent by Robert Pincus-Witten and Leon Hecht, New York

This group won the competition for the Darlington Memorial Fountain, Court House Square, Washington, D.C. A full-size replica, with gilding on the fillet and tassels of the drapery, once stood in the garden of Jennewein's architect friend Charles Louis Borie in Rydal, Pennsylvania (now in the collection of Brookgreen Gardens in South Carolina).

61.
Cupid and Crane
1926-27

Cast bronze, black patina, 21½″ x 14″ x 5″

Signed: *C.P. Jennewein/ c* (on top of base at center)

Foundry mark: *P.P.B. C° Munchen make in Germany* © (on right edge of base)

Collection of The Newark Museum. Gift of Mr. and Mrs. Felix Fuld 1927 (27.1152)

JACQUES (Chaim Jacob) LIPCHITZ
Born 1891, Druskieniki, Lithuania
Died 1973, Isle of Capri, Italy

As Henry R. Hope pointed out in his catalogue for the 1954 Museum of Modern Art retrospective, Lipchitz was essentially a modeler and "almost every sculpture he made was conceived in clay." Consequently, he was one of the few twentieth-century avant-garde sculptors to produce the majority of his works in bronze. By 1941, when he was forced to flee to the United States by the Nazis, Lipchitz was an established figure in the School of Paris, having worked there since 1908. Lipchitz experimented with Cubism between 1915 and 1919, and by the 1930s had arrived at a monumental and roughly modeled figurative idiom. The subjects of many of his large-scale sculptures were ancient Hebrew rites and sacrifices. After his New York City studio was destroyed by fire in 1952, the artist built a large studio in Hastings-on-Hudson, where he worked for the remainder of his prolific career.

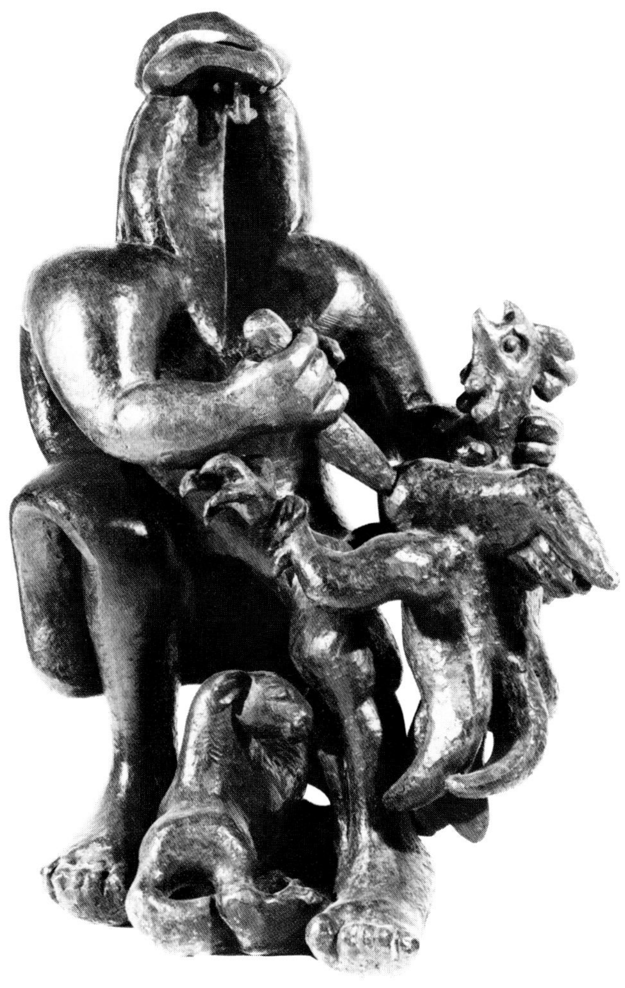

62.
Sacrifice, II
1948-52

Cast bronze, 49½″ x 33″ x 22″

Signed: *Lipchitz* (on back at left)

Lent by the Whitney Museum of American Art, New York. Puchase 1952 (52.27)

MAX KALISH
Born 1891, Cleveland, Ohio
Died 1945, New York City

Kalish studied at the Cleveland School of Art before going to New York to study at the National Academy of Design. In addition to his work at the Academy under A. Sterling Calder, Kalish received instruction from Herbert Adams, Isidore Konti and C.S. Pietro. The artist completed his formal studies in Paris at the Académie Colarossi and Ecole des Beaux-Arts, returning to the United States in 1913. Although Kalish executed several ideal subjects and portrait busts, his most consistent theme was the American worker. In 1932, the artist explained his preference: "it is my honest opinion that only by interpreting our own time, the Industrial Age, can we ever hope to produce a real epoch in art... In action an American worker is naturally graceful and makes a subject that even the Greeks might have envied." Unlike Social Realist painting of the 1930s and 1940s, Kalish's work never explored the socio-political implications of the Depression era workers he portrayed.

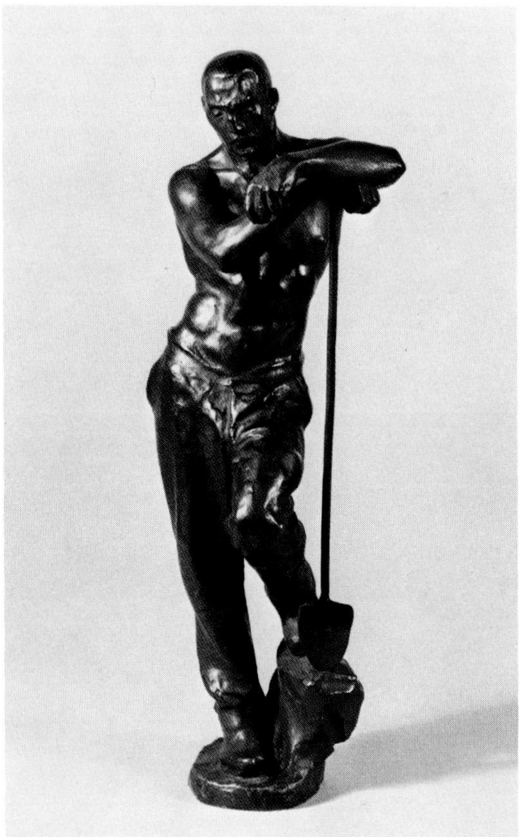

63.
Laborer at Rest
ca. 1924

Cast bronze, dark greenish brown patina, H. 15¼″

Signed: *M. Kalish.* (beneath pl foot)

Collection of The Newark Museum. Purchase 1927
J. Ackerman Coles Bequest Fund
(27.285)

REUBEN NAKIAN
Born 1897, College Point, N.Y.
Lives in Stamford, Conn.

Nakian moved with his family to New Jersey at the age of thirteen. He studied at the Art Students League, the Independent Art School and the Beaux Arts Institute of Design in New York. From 1916 to 1923, Nakian worked as an assistant to Paul Manship and shared a studio with Lachaise. Three years later, the artist had his first one-man exhibition at the Whitney Studio Club. Until the mid-1930s, he was best known for his small bronze sculptures of sleekly stylized animals. Under the influence of the painter Arshile Gorky, Nakian re-evaluated his style and produced little sculpture until the late 1940s, when he began working in terra-cotta on erotic themes from classical mythology. From 1945 to 1951, Nakian taught sculpture at the Newark School of Fine and Industrial Arts. Since the 1960s, the artist has again worked extensively in bronze, often in coarsely textured plasters that are cast in monumental size.

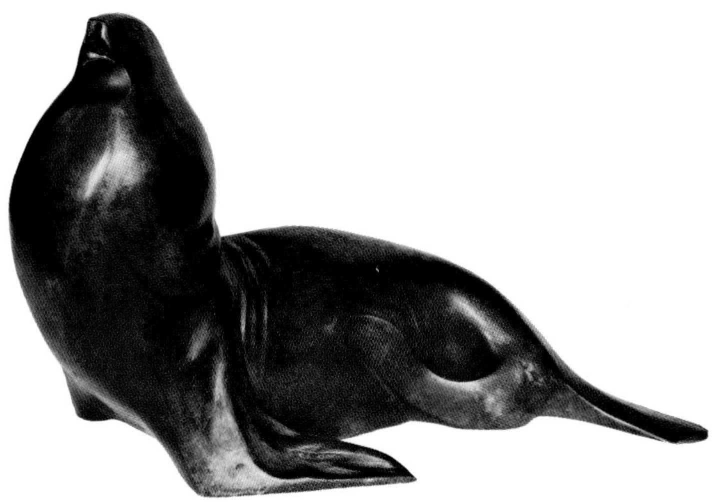

64.
Seal
1929

Cast bronze, golden brown patina, 8¾″ x 14″ x 7¾″

Signed: *-Nakian-* (on back at bottom)

Collection of The Newark Museum. Gift of Mrs. Edward S. Harkness 1931
(31.8)

DUNCAN FERGUSON
Born 1901, Shanghai, China
Died 1974

After graduating from Harvard University, Ferguson taught English and fine arts at colleges in the Boston area. He also began to study sculpture and became a student of Robert Laurent, with whom he maintained a lifelong friendship. Ferguson's first solo exhibition was in 1929 at the Downtown Gallery in New York. The artist worked primarily in bronze and taught for many years at Louisiana State University in Baton Rouge. Ferguson was married to the sculptor Alice Decker, also a student of Laurent.

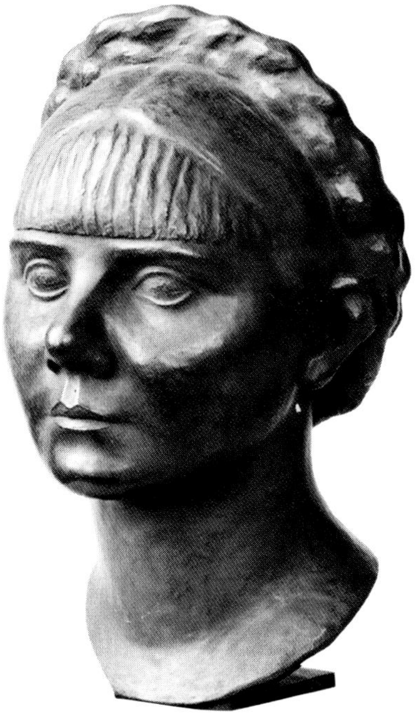

65.
Mimi *(Mrs. Robert Laurent)*
ca. 1928

Cast bronze, brown patina, H. 13¾″ on black marble base 6″ x 7″ x 5½″

Signed: *Ferguson* (on back at pr of neck)

Foundry mark: *Roman Bronze Works N-Y-* (on back)

Collection of The Newark Museum. Purchase 1928 General Fund (28.1762)

This piece is very similar to a head of the sitter modeled by Laurent (1924-27).

RICHMOND BARTHÉ
Born 1901, Bay St. Louis, Mo.
Lives in Pasadena, Calif.

Until he moved to Chicago in 1924 to study at the Art Institute, Barthé was a self-taught artist who supported himself with a variety of jobs. As a black man growing up in New Orleans, the artist was not able to attend any of the local art schools. Originally intending to be a painter, Barthé made his first sculptures—two clay heads—as visual aids to his painting at the suggestion of his teacher, Charles Schroeder. When these were exhibited at the Chicago Women's City Club in 1927, their success changed his career. Around 1930, Barthé moved to New York City, where he attended the Art Students League. Through his association with the Harmon Foundation and its traveling exhibitions, Barthé became one of the best-known black sculptors of the 1930s and 1940s. Although many of the artist's early sculpture dealt with specifically black subjects, he was also prolific as a portraitist. In 1947, the Grand Central Art Galleries in New York held a successful exhibition of his portrait busts of American and European actors. Not at home in the art world of New York, Barthé spent many years living abroad, first in Jamaica and then in Europe. At the age of eighty-three, Barthé is still active as a sculptor.

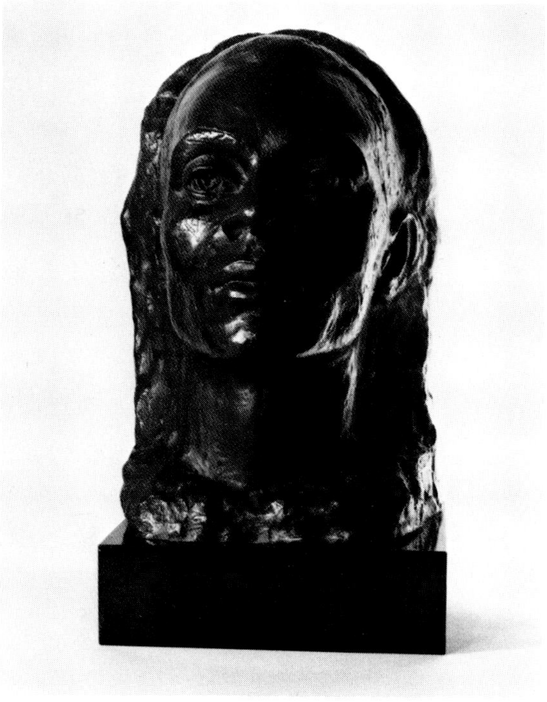

66.
The Negro Looks Ahead
Cast 1940

Cast bronze, dark green patina, 16″ x 10½″ x 10½″

No markings

Lent by The Schomburg Center for Research in Black Culture, The New York Public Library

SEYMOUR LIPTON
Born 1903, New York City
Lives in New York City

Entirely self-taught, Lipton did not turn seriously to sculpture until 1932. Before that time he had received his D.D.S. from Columbia University and worked as both a dentist and a sculptor. Although his earliest works were carved in wood, by the 1940s he was working primarily in metal, including bronze, lead and nickel-silver plated sheet steel. In his works from the 1940s and 1950s, Lipton often sought to expess what he called "a controlled organic dynamism." The twisting organic forms of these sculptures, along with their titles, further suggest the age-old mythic struggle between man and nature. Between 1940 and 1958, the artist taught at the New School for Social Research. He has had numerous one-man exhibitions and has created a large body of public sculptures, including *Archangel* (1964) in Avery Fisher Hall, Lincoln Center.

HERBERT FERBER (Silvers)
Born 1906, New York City
Lives in New York City

It was through his interest in anatomical drawing for dental classes at Columbia University that Ferber began to study art. He attended night classes in sculpture at the Beaux Arts Institute of Design in New York, where Leo Lantelli and Edward McCarten critiqued his work. After receiving his D.D.S. from Columbia in 1930, he established a teaching and dental practice and continued his work as an artist. Most of the artist's early work was carved directly in wood or stone in a manner greatly influenced by William Zorach and Robert Laurent. Ferber first began to work consistently in bronze in the 1940s. During this decade he absorbed a variety of influences, including the work of Henry Moore, Julio Gonzales and Joan Miro, arriving at style that combined biomorphic imagery of a deeply personal and emotive kind with the skeletal framework of Constructivist sculpture. For his 1951 sculpture for the facade of the B'nai Israel Synagogue in Millburn, New Jersey, Ferber developed these forms into a more open and fluid composition on the theme *"...and the bush was not consumed."* A few years later the geometric underpinning of the artist's earlier work began to reassert itself in a series of calligraphic forms, a development that has continued through his present work. During the 1950s and 1960s, Ferber was closely associated with other painters and sculptors of the Abstract Expressionist group. He continues to work in a variety of materials and exhibits regularly in New York.

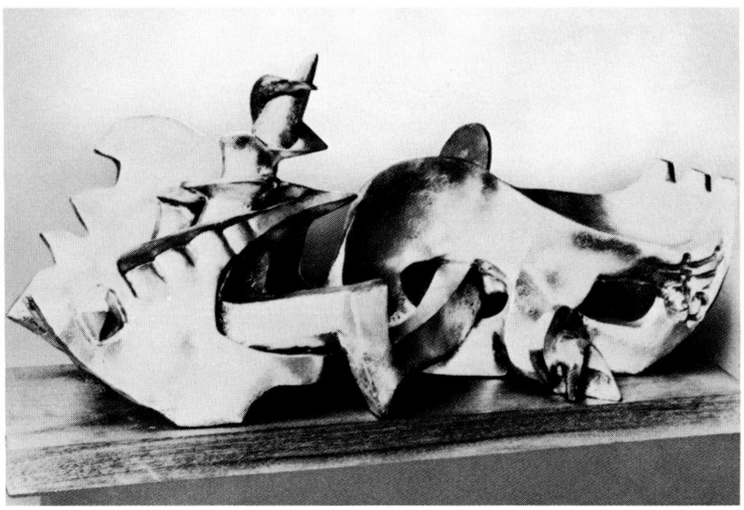

67.
Argosy
1948

Cast bronze, golden brown patina, 9¾″ x 28½″ x 14″

Signed: ©/*Lipton '48* (on front near bottom right)

Lent by the Grey Art Gallery and Study Center. New York University Art Collection. Anonymous Gift, 1966

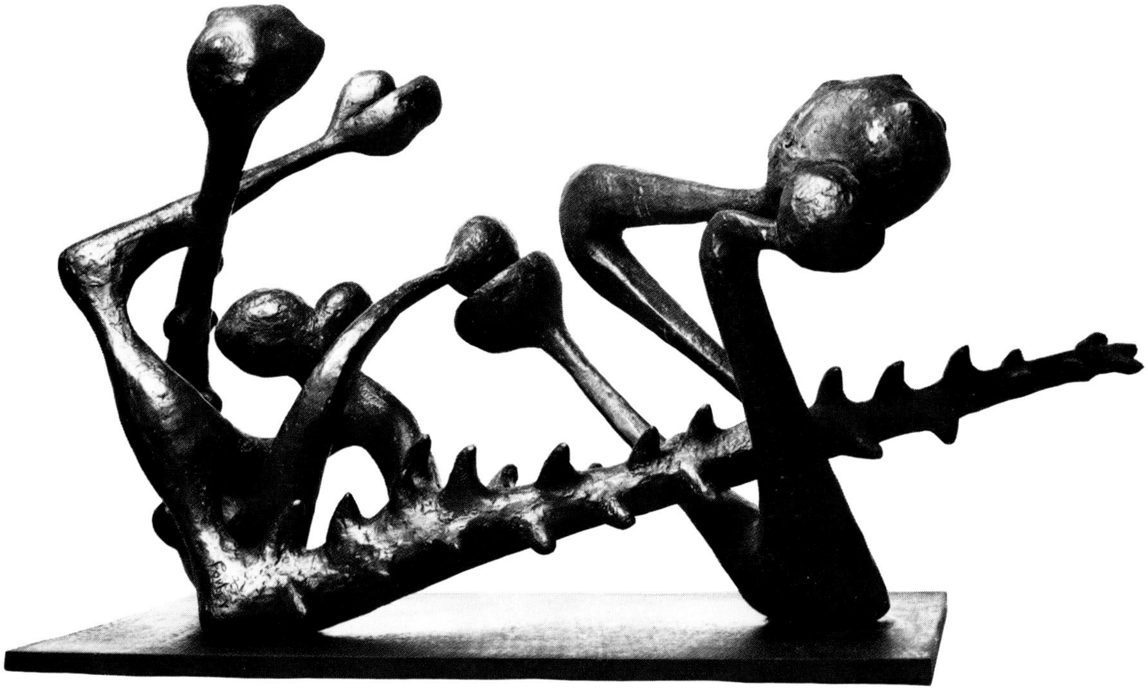

68.
Hazardous Encounter
1947

Cast bronze, dark brown patina, 21½" x 37" x 13"; oak base, 2" x 28½" x 11½"

Signed and dated: *Ferber 47* (on front near bottom)

Lent courtesy of M. Knoedler & Co., Inc., New York

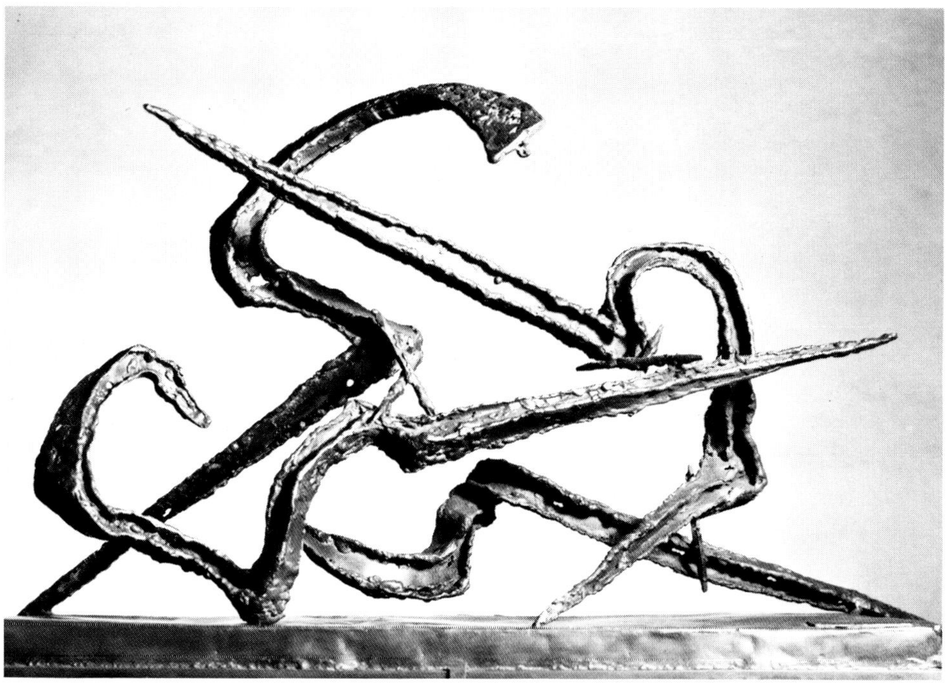

69.
Calligraph in Four Parts
1957

Welded bronze, 14⅞" x 25½" x 11"

Signed and dated: *Ferber 57* (on rectangular plate welded to top of base)

Collection of The Newark Museum. Purchase 1960 John J. O'Neill Bequest Fund (60.579)

MATTHEW SAFFERSON
Born 1909, London, England
Lives in Brooklyn, N.Y.

Safferson immigrated to the United States in 1923. As a teenager he took art classes at the Hebrew Educational Alliance in New York. The artist then attended Cooper Union and the Leonardo Da Vinci Art School. Safferson completed his sculpture training as an apprentice to Leo Friedlander. Unlike his teacher, Safferson embraced the theories of modern sculpture and created abstract sculptures composed of geometric forms. Although little is known of his subsequent work, during the 1940s the artist worked for the W.P.A., New York Project.

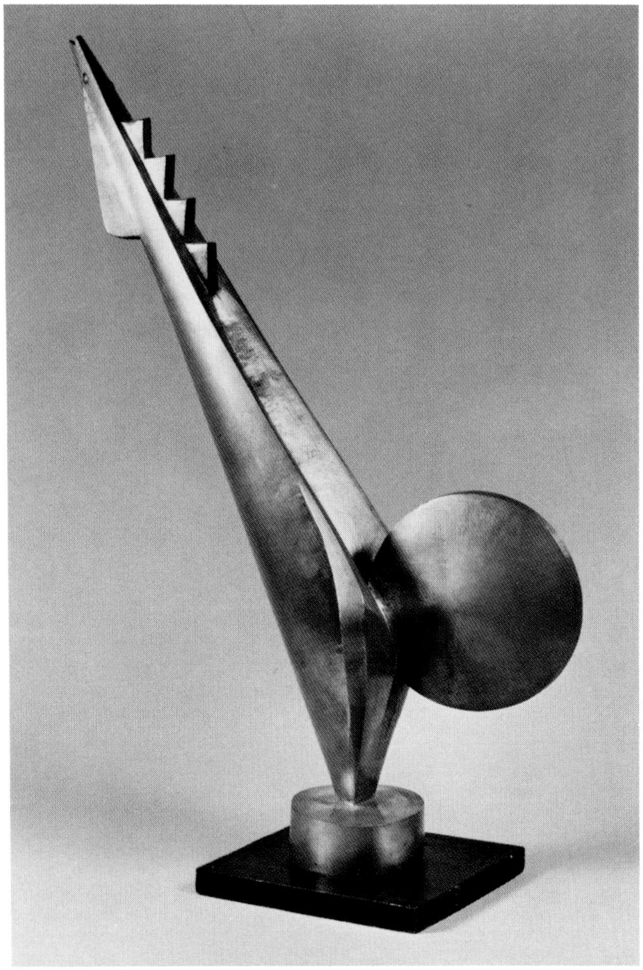

70.
Rooster
ca. 1940

Polished bronze on black painted wood base, 26½" x 17¾" x 5⅝"

No markings

Collection of The Newark Museum. Allocated by the WPA Federal Art Project 1943 (43.151)

PETER J. GRIPPE
Born 1912, Buffalo, N.Y.
Lives in Orient, N.Y.

Grippe received his early training at the Albright Art School (1923-25) and the Art Institute of Buffalo (1929-35) with the painter Edwin Dickinson. From 1944 to 1948, he worked at William Stanley Hayter's Atelier 17 in New York, where he served as director from 1951 to 1954. During these years he worked frequently as a printmaker. Like most artists of his generation, Grippe was greatly influenced by both Surrealism and the later manifestations of Cubist theories. Like Herbert Ferber, he often enclosed his forms in an open grid structure. Grippe has been active as a teacher, serving on the faculties of Black Mountain College and Brandeis University.

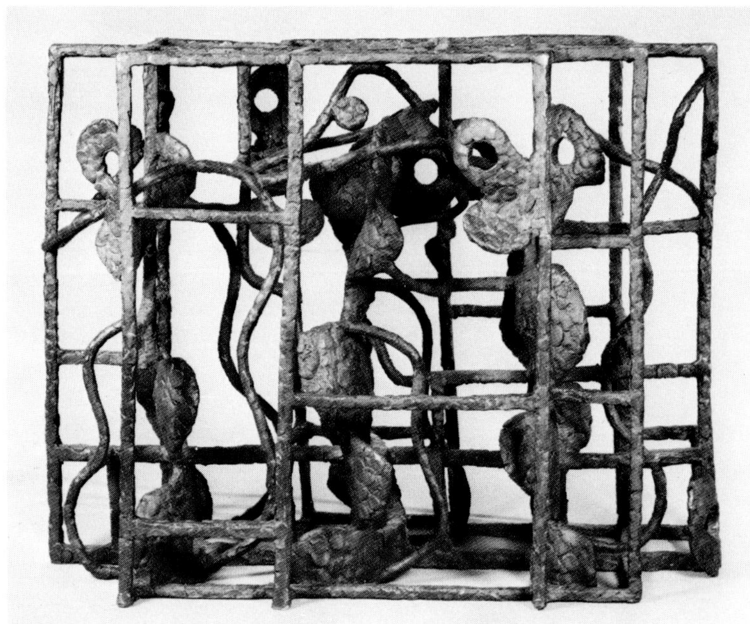

71.
Symbolic Figure No. 4
(Theseus in the Labyrinth)
1946

Cast bronze, green patina, 17" x 11" x 22"

Signed: *G4* (on base)

No foundry mark (cast at Modern Art Foundry, N.Y.)

Collection of The Newark Museum. Gift of Willard Gallery 1962 (62.146)

TONY SMITH
Born 1912, South Orange, N.J.
Died 1980, South Orange, N.J.

Smith was perhaps the most important and influential of the Minimal sculptors of the 1960s and 1970s. Like other artists associated with this movement, Smith prefered to use common industrial materials and processes and emphasized a clear projection of simple formal relationships. During the 1930s, while working in his family's waterworks foundry, he attended the Art Students League, where he studied with George Grosz and Vaclav Vytlacil. In 1937, Smith enrolled in the New Bauhaus (Chicago) but was disappointed in how little benefit he derived from the school. The next two years were spent as an apprentice to the architect Frank Lloyd Wright. Until giving up his practice in the 1960s, Smith taught and worked as an architect. His transition to sculpture is most clearly explained in the "Account of Career" he wrote for the catalogue of an exhibition held at The Newark Museum in 1970:

"During an interlude in the above period, 1953-55, I lived in Germany and devoted myself to projects in painting, sculpture and architecture. Except for some small paintings, drawings and notebooks, there is nothing to show from this time; for me, however, it was the most fruitful period of work. In about 1962, I became intolerant of the capriciousness of clients and I gave up building. I had some steel boxes made and placed them around in my yard. I don't know exactly what my intentions were. Later, I began to develop some forms based on various types of space frames. These were clearly intended as sculptural expressions."

Smith usually constructed his sculptures in plywood and had them commercially fabricated in steel or, less frequently, in bronze. This method of working, with its denial of the artist's "touch," represented a profound break with the traditional use of bronze as a personally expressive material.

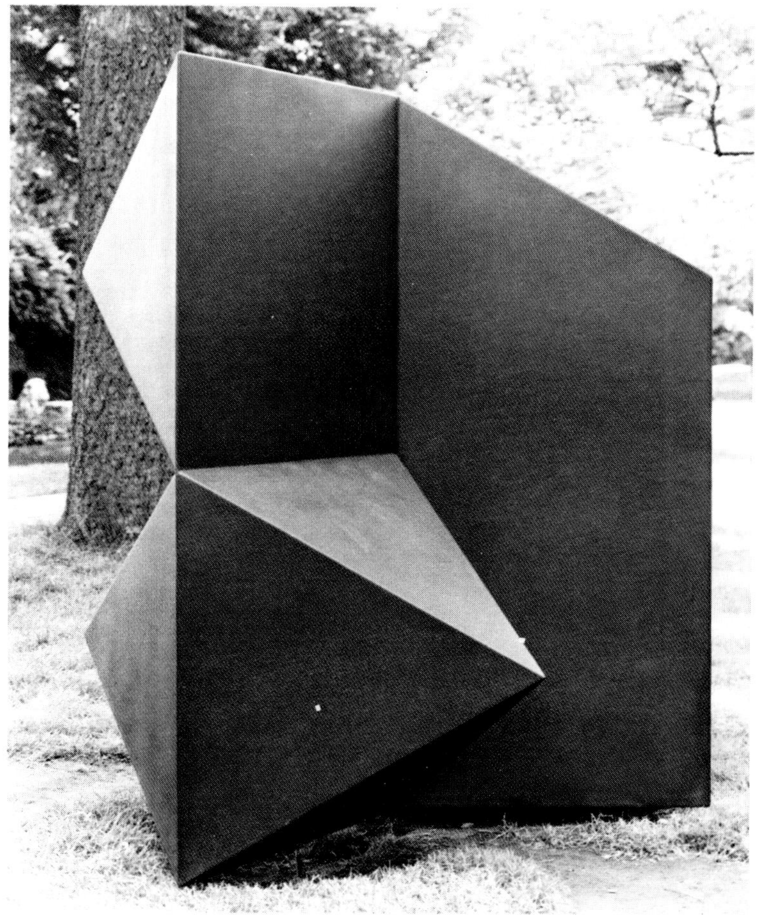

72.
For J.C.
1969-70; fabricated 1975

Welded bronze, dark brown/black patina, 80″ x 80″ x 57″ (rhomboidal)

No markings (fabricated by Industrial Welding Company, Newark)

Collection of The Newark Museum. Purchase 1975 with a grant from the National Endowment for the Arts and funds from the Prudential Insurance Company of America (75.232)

Fabricated especially for The Newark Museum, this is the first of an intended edition of six. The Museum also owns two small maquettes for the piece. "J.C." refers to the French cubist painter Jean Charlot (1898-1980).

DAVID HARE
Born 1917, New York City
Lives in New York City

Although he created his first sculptures while in high school, as a young man Hare was primarily interested in chemistry and medicine and never formally studied art. In the 1930s, Hare worked as a photographer, manipulating his images by the process of "heatage" developed by the Surrealist Raoul Ubac. In 1942 he exhibited in the landmark *The First Papers of Surrealism* exhibition and helped found and edit the Surrealist magazine *VVV*. In the same year he began to sculpt, creating works in a distinctly Surrealist idiom. These sculptures, usually created in plaster for casting in bronze, were characterized by the historian Robert Goldwater as "neither abstract, nor stylized, nor representational . . . images that intensify, rather than reduce the natural form." In 1951, Hare began to work in welded metal, using mixtures of steel, lead, bronze and copper.

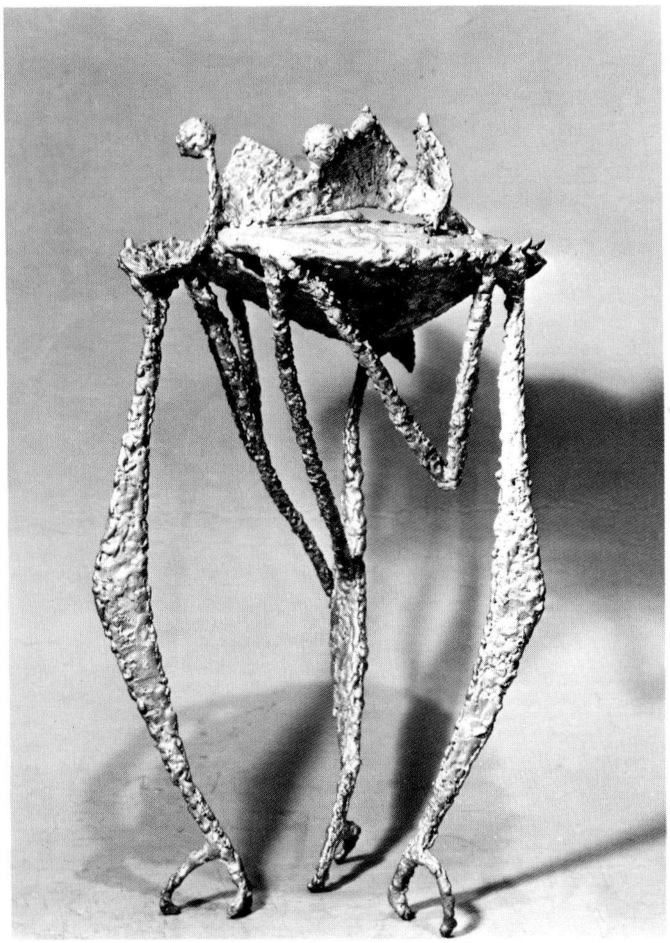

73.
Crab
1951

Welded bronze, H. 23¼", on three points 9" x 7½" x 7¼" apart

Lent by The Museum of Modern Art, New York. Purchase

HAROLD TOVISH
Born 1921, New York City
Lives in Boston, Mass.

Tovish received his first art instruction in art classes run by the W.P.A. during the late 1930s. After attending Columbia University (1940-43), the artist traveled to Paris, where he studied at Ossip Zadkine's School of Sculpture and at the Académie de la Grande Chaumiere. Since the early 1950s, Tovish has worked in the Boston area, teaching at the School of the Boston Museum of Fine Arts, Massachusetts Institute of Technology and Boston University. In 1965, he was sculptor in residence at the American Academy in Rome.

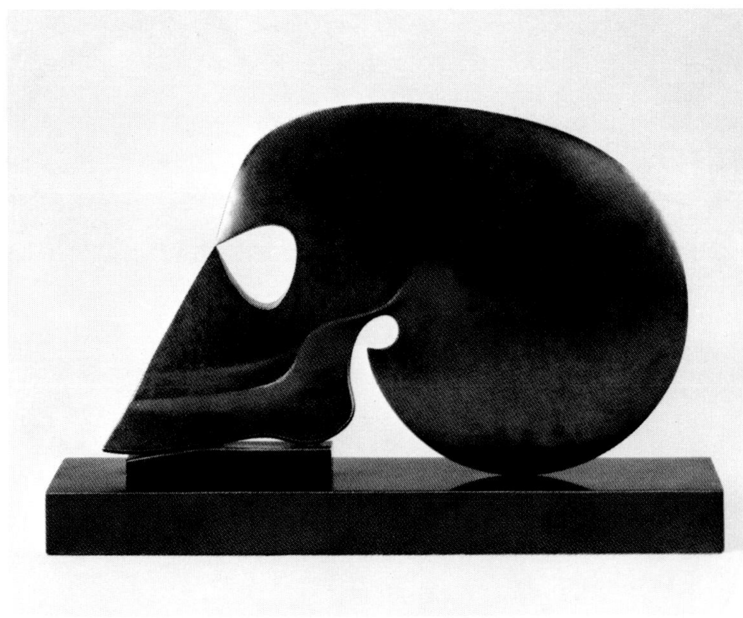

74.
Ceremonial Axe Head
1975

Cast bronze, golden brown patina, 7¼" x 11¼" x ⅝"; base, 1½" x 13" x 2¾"

No markings

Private Collection

LEONARD BASKIN

Born 1922, New Brunswick, N.J.
Lives in Northampton, Mass.

The son of a rabbi, Baskin is an extremely well-educated artist. He grew up and was educated at Yeshiva in Brooklyn and as a young man he studied with Maurice Glickman who taught at the Educational Alliance in New York City. Later he attended the New York University School of Architecture and Allied Arts (1939-41) and briefly, Yale University. After serving in the Navy during World War II, Baskin returned to New York. The artist was awarded a Tiffany Foundation Fellowship in 1947 and, two years later, completed his bachelor's degree at the New School for Social Research, New York. In 1950, Baskin returned briefly to Europe, attending art schools in Paris and Florence under the G.I. Bill. The artist has worked in a variety of media but is perhaps best known for his large-scale sculptures in bronze and carved wood. He was for many years on the faculty of Smith College.

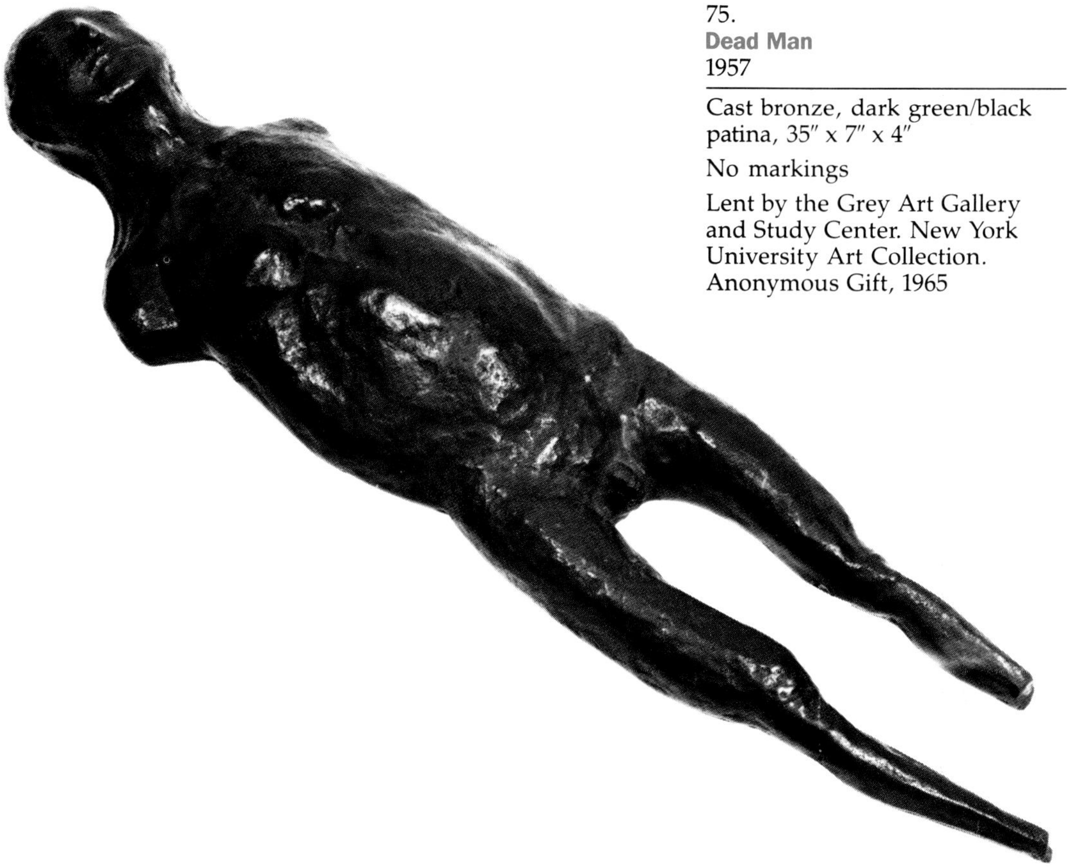

75.
Dead Man
1957

Cast bronze, dark green/black patina, 35″ x 7″ x 4″

No markings

Lent by the Grey Art Gallery and Study Center. New York University Art Collection. Anonymous Gift, 1965

GEORGE SEGAL
Born 1924, New York City
Lives in New Brunswick, N.J.

Moving with his family to New Brunswick in 1940, Segal attended Rutgers University and Pratt Institute before receiving his bachelor's degree in art education from New York University in 1950. He taught for several years in New Jersey public schools and owned a chicken farm, where he converted one of the buildings into a studio. Although he worked during the early 1950s for several years as a painter, Segal's first critical attention came as a result of the plaster, wire and burlap bas-reliefs of human figures that he exhibited at the Hansa Gallery, New York, in 1958. By 1965, when he began to exhibit regularly at the Janis Gallery in New York, his plaster sculptures had evolved into fully three-dimensional figures, often combined with real objects, such as tables, chairs or doorways, to create emotionally charged "situations" from life. The artist first began to cast his figures in bronze in 1975, when he received a commission from the Federal government for an outdoor sculpture in Buffalo, New York (*Restaurant*, 1975-76, Federal Building, West Huron Street). The figures for this piece were patinated in a traditional dark green, but a few years later the Johnson Atelier, in Princeton, New Jersey, developed a white patina for Segal's outdoor sculptures that approximates the look of plaster. The artist continues to create his sculptures for both media.

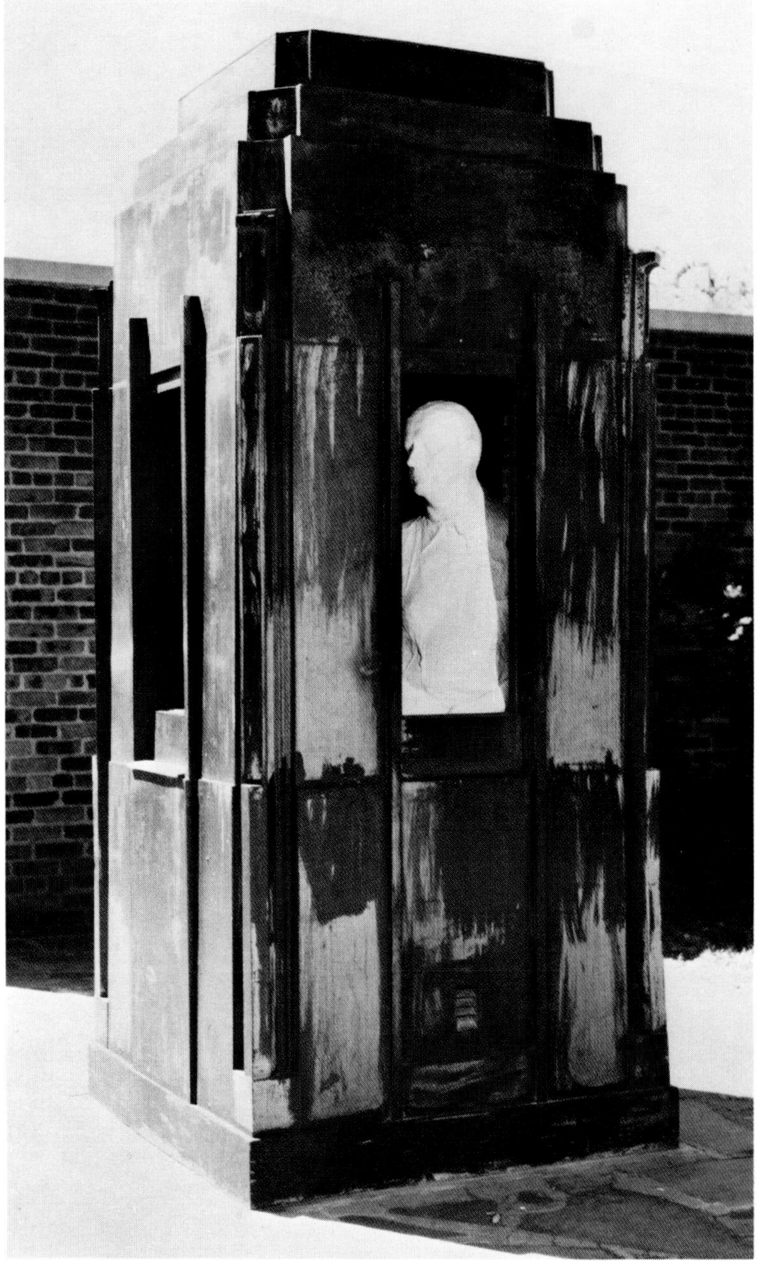

76.
Toll Booth Collector
1980

Cast bronze, white patina, H. 74″

No markings (cast at Johnson Atelier, Princeton, New Jersey)

Collection of The Newark Museum. Purchased with a grant from the National Endowment for the Arts and Visitors' Contributions 1980
(80.5)

Segal created this figure to be placed inside a toll booth from the Holland Tunnel given to the Museum in 1977 by the Port Authority of New York and New Jersey. It was cast from a direct mold made of the Museum's Director, Samuel C. Miller.

WILLIAM KING
Born 1925, Jacksonville, Fla.
Lives in New York City

Since he first began exhibiting his sculpture in the late 1940s, King has worked in a broad range of materials, including wood, bronze, aluminum, terra-cotta and fabric. He studied engineering at the University of Florida before moving to New York to attend the Cooper Union and the Brooklyn Museum Art School. In 1949, the artist received a Fulbright Fellowship to travel to Rome, where he studied at the Academia dei Belli Artes. Although the materials of his sculptures may vary (and a great deal of it has been created in bronze), the sense of wit and humor remains constant. King's tall, lanky figures often display an inventiveness of form and construction that once prompted the critic John Canaday to call them "not only clever, but witty, which means that perceptive comment is combined with ingenious presentation."

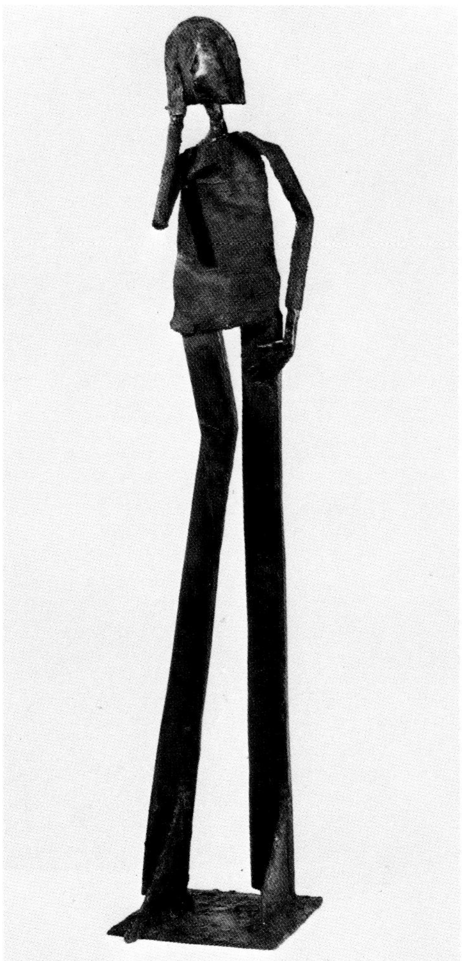

77.
St. Cecilia
1981

Cast bronze, H. 38″

No markings

Lent courtesy of Terry Dintenfass Gallery, N.Y.

Saint Cecilia is the patroness of music.

ISAAC WITKIN
Born 1936, Johannesburg, South Africa
Lives in Bennington, Vermont

In 1956, Witkin moved to England, where he studied with Anthony Caro at the St. Martin's School of Art, London. After finishing his formal studies in 1960, he spent three years working as an assistant to Henry Moore. During the 1960s, Witkin established his career as a Minimalist sculptor. His works were most often created of welded steel and intended for large outdoor spaces. Like other artists of his generation, he was interested in using industrial materials that could be worked into abstract images that had a pared down, almost muscular, strength. In 1965, the artist was invited to teach at Bennington College in Vermont, where he continues to work. The bronze sculptures that Witkin has created since the early 1980s are a radical departure from his earlier work. Although the artist continues his interest in balancing shapes and volumes, the newer works are much more organic and spontaneous in feeling. For most of these sculptures, Witkin pours the liquid bronze directly into open shallow sand molds that have pre-established framed contours. The individual pieces are then welded together and patinated. This process retains the individual gesture of each pour and heightens the sense of the artist's "touch."

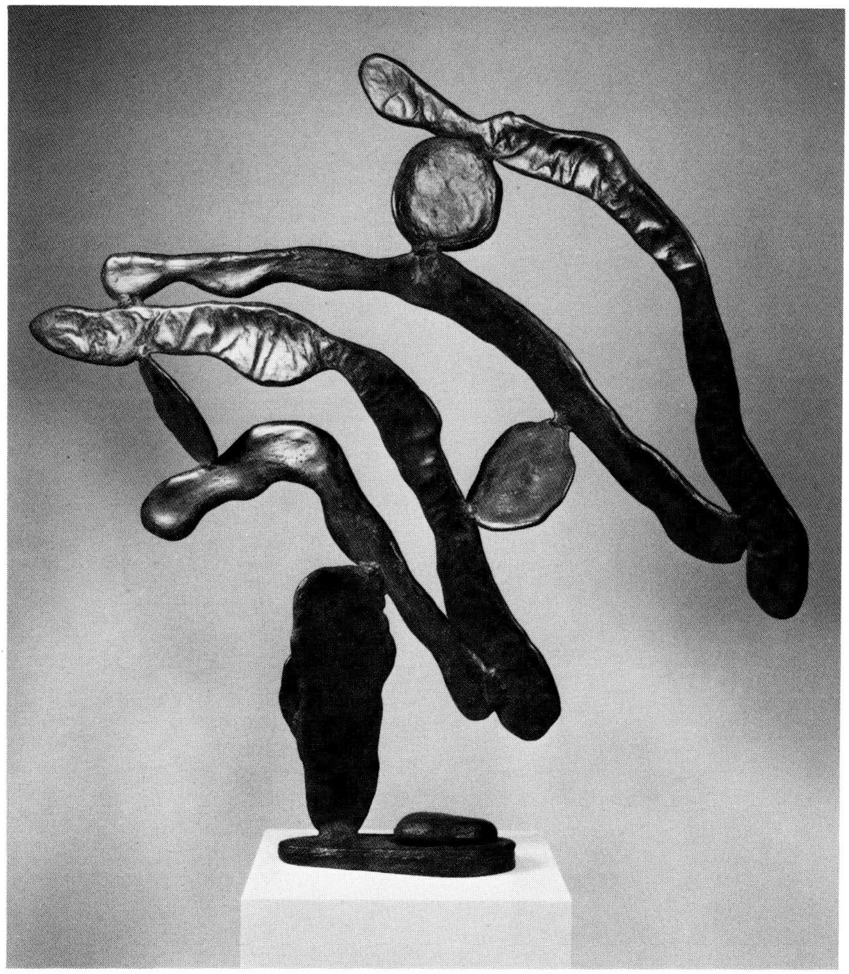

78.
Firebird II
1984

Cast bronze, pale red-pink patina, 54″ x 57″ x 6″

No markings (cast at Johnson Atelier, Princeton, N.J.)

Lent courtesy of Hirschl & Adler Modern, N.Y.

RED (Charles) GROOMS
Born 1937, Nashville, Tenn.
Lives in New York City

Grooms studied at the School of the Art Institute of Chicago and the George Peabody College for Teachers, Nashville, before arriving in New York City at the age of nineteen. The following year, 1957, he attended Hans Hoffman's School of Fine Arts in Provincetown, Massachusetts. Grooms quickly became one of the most exuberant and active artists of the New York art scene during the 1960s. He created happenings, videos and films, paintings and mixed media constructions. Today the artist is best known for his humorous, out-scale sculpture environments, such as *The Discount Store* (1970) and *Ruckus Manhattan* (1976). Grooms' use of bronze began in 1979, while at the University of New Mexico in Albuquerque, at the invitation of his friend Paul Suttman. The resulting series of comical Western subjects, along with four new football pieces, were cast the following year at Southern Illinois University, where Grooms again taught with Suttman. Each of the pieces was painted by the artist in the manner of his other sculptures. In 1981, Grooms said of his new painted bronzes, "It looks just like my regular stuff, but it's for the ages.... It turns out to be easier to work with than less durable materials. I don't want to concentrate on it, but it's nice to surprise yourself. I get a big kick out of using new material, because in it you can regenerate your subject matter." In this spirit, the artist has continued to work with bronze.

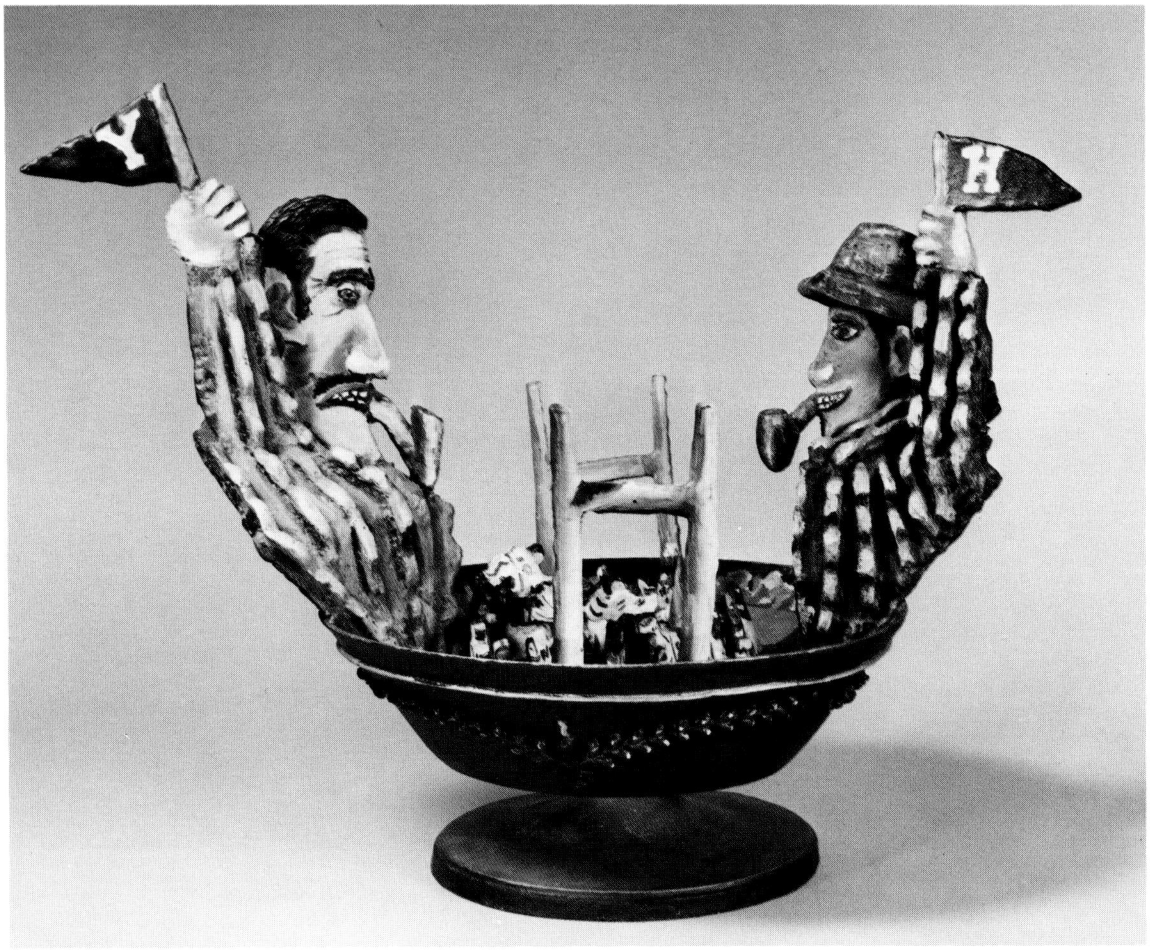

79.
Harvard Yale Bowl
1980

Cast bronze, painted,
18¾″ x 25″ x 14¼″

Signed: *Red Grooms* (in red paint on base of bowl at right)

Numbered: *3/6* (in red paint on base of bowl at left)

Lent courtesy of Marlborough Galleries, Inc., New York

ROBERT GRAHAM
Born 1938, Mexico City
Lives in Venice, Calif.

After moving to the United States in 1950, Graham grew up in California, where he attended San Jose State College and the San Francisco Art Institute. In the mid-1960s, he gained critical recognition with his sculptures of small, delicately modeled wax figures, placed in plexiglass environments. After living and working in London in 1969 and 1970, the artist returned to Los Angeles and began casting his sculptures in bronze. Although these new figures often remained small, or at least less than life size, they were no longer encased in plexiglass nor as patently illusionistic. Usually hairless and blankeyed, his female nudes began to appear as fragments on top of box-like structures or columns. Although extremely lifelike in detail, Graham's figures have an unsettling appearance that has been described as "hallucinatory." The artist has continued to work primarily with the human figure, sometimes painting the surfaces of his bronzes or adding small illusionistic details in other materials. Graham's studio includes a large foundry where, with the help of assistants, he casts his own work.

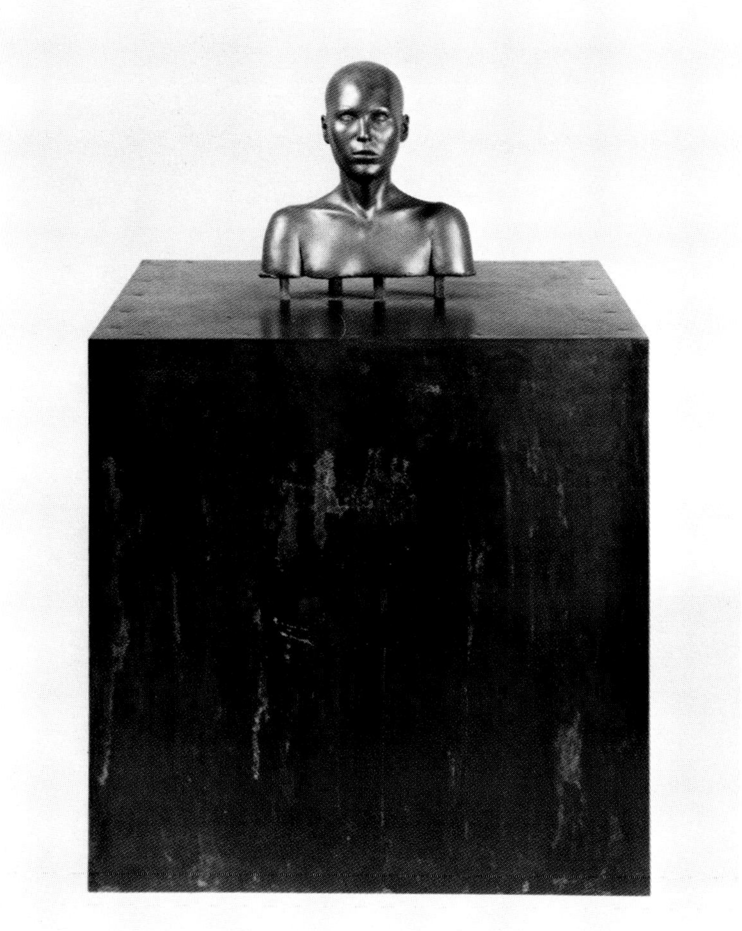

80.
Single Head
1973

Cast bronze head, golden brown patina, on fabricated bronze base, 17½" x 12" x 12" (overall)

No markings

Lent anonymously. Promised gift to The Newark Museum

MARJORIE STRIDER
Born in Guthrie, Okla.
Lives in New York City

Strider has worked with a wide variety of materials and has often been linked with the Pop Art movement of the 1960s and 1970s. However, like those of Red Grooms, her sculptures have a droll sense of humor and an out-of-scale proportion that keeps them from fitting tidily into any contemporary movement. Growing up in Oklahoma, Strider attended the Kansas City Art Institute and then moved to New York City. After her work was included in a group show at the Pace Gallery, New York, in 1964, Strider became best known for her brightly painted wall reliefs which appeared to explode out toward the viewer. During the 1970s, the artist created a series of painted sculptures of fruits and vegetables with highly detailed surfaces that explored formal questions of space and scale. It was while working on these pieces that Strider first began to cast her sculpture in bronze. In addition to her painting, sculpting and printmaking, the artist has also created many films and performance pieces. Strider has been a member of the faculty of School of Visual Arts, New York, since 1969 and taught painting classes at The Newark Museum from 1972 to 1974.

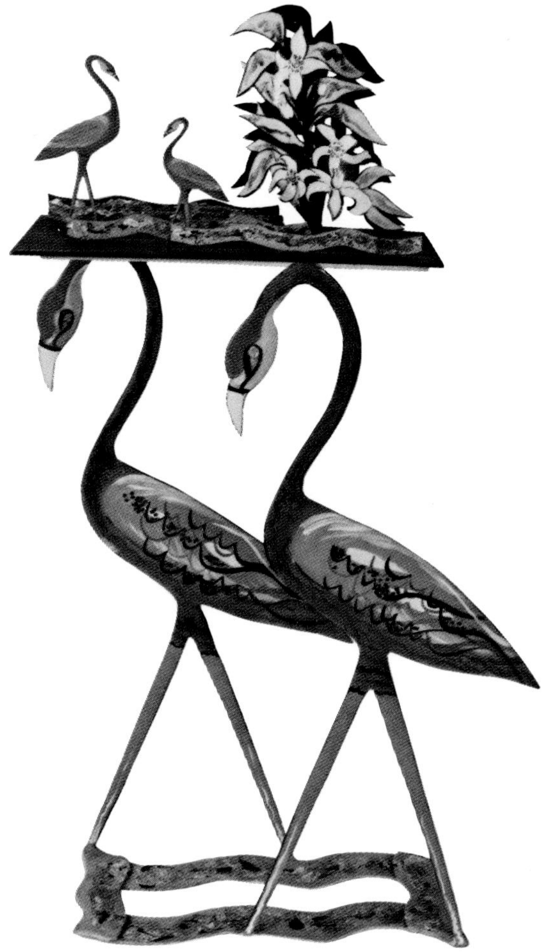

82.
Florida II
1981

Bronze and aluminum, painted, applied sequins and rhinestones, 50″ x 28″ x 10″

Signed and dated: *M. Strider* © *81* (in paint on top of base at back)

Lent courtesy of Bernice Steinbaum Gallery

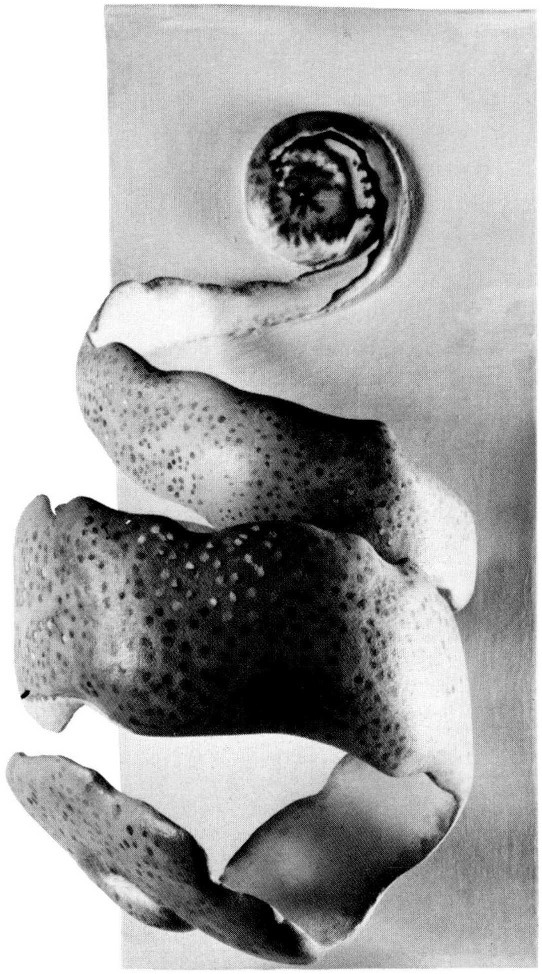

81.
Peel I
1975

Cast bronze, painted, 15″ x 8″ x 9″

Signed and dated: © *Strider* (in paint on back)

Lent by Berta Walker

NUMISMATICS

ED REITER

ɔld Pandas and Penny Boards

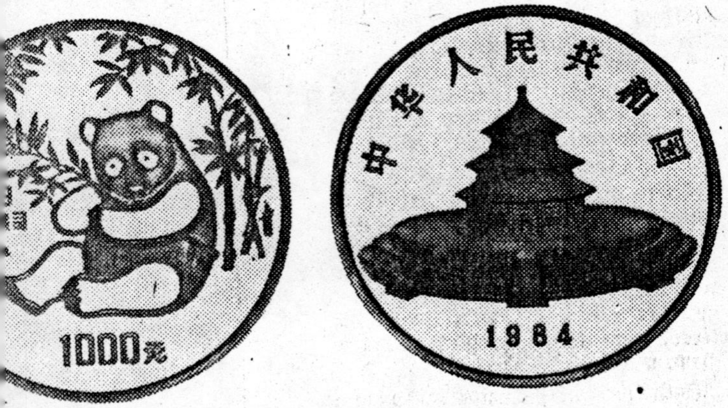

ɔld Panda coin from Republic of China contains 12 troy ounces.

apparent that people who d them simply as diversions oming intrigued with numis-

Mr. Post's creation has the face of the hobby and o pave the way for a far market. Without exaggera- coin or "penny" board revo- d numismatics and ushered den age for the hobby. Ken- Bressett, editor of Whitman's selling guide books on United coins, the "Blue Book" and ook," accepts this conclusion reservation.

n this began," Mr. Bressett d, "the board was simply as a leisure-time activity, ke Whitman's jigsaw puzzles aying cards. Coin collectors l it, because it was such a con- means of storage. Then it be- case of the tail wagging the

United States Coins," first appeared in 1942 and became an immediate success. Now in its 42nd edition, the book gives wholesale values for coins — those that a dealer might pay to obtain them from a collector. The Red Book — "A Guide Book of United States Coins" — first appeared in 1947 and was recently released in its 38th edition. It lists retail values — those a collector might have to pay. Between them, the books have sold upwards of 20 million copies which ranks them among the all-time nonfiction best-sellers.

It is clear to Mr. Yeoman that the Blue Book and Red Book and the new collectors they have spawned never would have come into being without the introduction of the coin board. "The system of collecting by dates and mint marks made all the difference in the world," he said. "The idea of filling holes one by one until the col-

Head nickel, the Buffalo nickel, the Barber dime, the Mercury dime and the Standing Liberty quarter.

The boards for the Barber quarters and dimes described them as "Morgan-Liberty Head" coins — an obvious misnomer since both (as well as the Barber half dollar) were designed by Charles E. Barber, not George T. Morgan when both were Mint engravers. The Standing Liberty quarter board provided holes from 1916 to 1936 — even though the series ended in 1930. The absence of half dollar and silver dollar boards may puzzle present-day collectors, since both denominations are now extremely popular. However, in the mid-1930's, those coins represented substantial sums of money and Whitman perceived that not many people would be in a position to set them aside.

At present, the Whitman Coin Products Division of Western Publishing Company, is producing 23 coin folders and 16 "Classic" Bookshelf albums. Suggested retail price for the folders is $1.59 apiece. The albums range from $7.95 to $12.95.

Demand for folders has been dampened in recent years by the sharp drop in the number of date-mint combinations available in circulation. There has also been widespread hobby apathy toward the cupronickel coins of the last two decades, yet annual sales remain in the hundreds of thousands.

Cameo Proof Book

A comprehensive book on modern U.S. cameo proof coins has been written by Val J. Webb of Atlanta, Ga. Entitled "Cameo Proofs, 1950-1964,"

ART VIEW

Continued from Page 29

skull-like object of sensuously rounded geometric forms with a gruesome tooth-lined maw — both works in conventional bronze patina. Other artists, some making use of give bronze any color and the semblance of other material, have made innovative contributions. Take, for example, Bryan Hunt's "Cloak of Lorenzo," a flattened, rearing form of very agitated surface, patinated in deep black, that suggests at once a waterfall, a torso and a very broad Abstract Expressionist brushstroke. Or Nancy Graves's "Lacinate," a wild, exuberantly painted assemblage of plant and floral materials.

In "Wave," Lynda Benglis has worked bronze into a frozen explosion of water, patinated in green, that harks back to the wave of horses by Anna Hyatt Huntington, and Joel Shapiro has cast in golden bronze a simple totemic structure of wooden planks that obliquely refers to the human figure. Not to be missed is another garden piece, an actual bronze toll booth of Art Deco inspiration, donated to the museum by the Port Authority. In it, George Segal has placed a pensive toll collector, whose figure has been cast in bronze from that of the museum's director, Samuel Miller.

The show, handsomely installed by David Palmer and Danny Schnur of the museum's exhibitions department, is smashing in its own right, and also interesting as a complement to the Whitney Museum's current exhibition, "The Third Dimension: Sculpture of the New York School," which contains a number of bronzes by the Abstract Expressionist sculptors mentioned above. And the Newark Museum, at 49 Washington Street, is quickly reachable by PATH train or Amtrak from Pennsylvania Station. ■

THE NEW YORK TIMES
SUNDAY, DECEMBER 23, 1984

Art

"The uses of bronze can run to the pompous as well as the sublime." (Grace Glueck)

ART VIEW

GRACE GLUECK

Bronze in the Hands Of American Sculptors

There is nothing flash in the pan about bronze. From antiquity, it's been regarded as a serious, monumental metal, a metaphor for permanence, employed for art and ritual objects, weapons and statues of achievers. But in this young country, it didn't come into widespread use for sculpture until the middle of the 19th century, when we had developed enough sense of history to want more lasting monuments, and achieved the bronze technology to make them. Turning from carving in wood and marble, American sculptors learned to model in this durable alloy of copper and tin, and were eventually able to realize as skillfully as their European counterparts its richly expressive potential.

To celebrate their work, the Newark Museum has mounted "American Bronze Sculpture" (through Feb. 3), the first large-scale museum show to trace the sculptural career of bronze in this country. Its 88 pieces, the work of 57 artists, range from an 1852 warts-and-all bust of Henry Clay by Henry Kirke Brown to contemporary abstract fantasies by Nancy Graves and Bryan Hunt. It's a splendid presentation — revealing that the uses of bronze can

That bronze is in another revival stage is left in no doubt by the show.

run to the pompous as well as the sublime — and a good part of it comes from Newark's own fine collection of small-scale bronzes, discovered in basement storage by Gary A. Reynolds, the museum's new curator. Aware of the current revival of sculptural interest in the medium, he put them together with contemporary work — some also owned by Newark — et voila! a show was born.

The first successful monumental bronze made in this country, Clark Mills's statue of Andrew Jackson on a rearing horse, is still standing in Washington's Lafayette Square. This engaging piece, naive but peppy, was unveiled in 1853, after the sculptor had invested years of trial and error in its completion, all for the lack of a commercial foundry skilled at casting large-scale sculpture. Meanwhile, needing facilities for his own work, Henry Kirke Brown, back in the States after four years of study in Rome, built a foundry in his Brooklyn studio in 1848. He could cast small pieces there, such as the bust of Henry Clay. But for his first monumental commission, the statue of George Washington on horseback that stands in Manhattan's Union Square, he turned to the Ames Manufacturing Company, a weapons foundry in Massachusetts. The statue, well conceived and ably executed, was dedicated on July 4, 1856. It not only launched Brown's career; it propelled Ames and other foundries, spurred by growing demand, into the business of casting bronze sculpture.

Until the late 1890's, Mr. Reynolds points out in his catalogue introduction, American foundries used only the relatively crude procedure of sand-casting, which made difficult the reproduction of intricate sculptural details. Despite that problem, many of the early pieces show strong and lively handling, among them Olin Levi Warner's portrait bust of the musician Maud Morgan, 1880, notable for its poignant, naturalistic rendering of a soulful face. (Contrast the pleasant kitsch of Harriet Hosmer's "Hands of Robert and Elizabeth Barrett Browning," 1853, a pair of disembodied hands cast from the poets' own, in a possessive clasp.) Then, in 1897, the ancient lost-wax process, revived in Europe by the mid-19th century, was brought here by two Italian immigrants, Riccardo Bertelli and Giuseppe Moretti, who set up the first art foundry to do casting by this method.

• • •

A big improvement over sand-casting, the method provided for the creation of a flexible mold that took exactly the imprint of the artist's model. Any surface modeling or detail could be faithfully reproduced, even the deep undercutting that sand-casting didn't encourage. What's more, unlike sand-casting, it allowed for the molding of sculptures in one piece, and since the mold was reusable, sculpture could be produced in multiple units. The new method coincided with a new approach to sculpture. By the 1880's, young American sculptors had virtually abandoned Italy — the stronghold, for nearly half-a-century, of pure white marble Neo-Classicism — for Paris, where a shift not only in style but medium had surfaced. Modeling had succeeded carving as the order of the day, taking into account the fluid forms and expressive surfaces obtainable — most notably by Rodin — with bronze. The American sculptors, among them two who would later achieve real fame, Augustus Saint-Gaudens and Daniel Chester French, came home to spread the new Parisian gospel.

St. Gaudens is represesented here by a sturdy model for a monument to Lincoln, in which the ornate chair Lincoln fronts in a rhetorical pose is given the same weighty presence as his figure. In another St. Gaudens mood is a very Beaux Arts winged goddess, voluminously draped, a symbol of high thoughts and elevated feelings, who holds up a tablet reading "Amor.Caritas." Though French is best remembered for his magnificent Lincoln Memorial in Washington, D.C., carved of white marble, he, too, made a Lincoln monument in bronze, a less pose-y standing figure, whose cast-down head and awkwardly placed feet suggest the shy, ungainly stance of the late Gary Cooper. But his "Spirit of Life," 1914, a lovely fountain figure of a winged nymph with arms upstretched, is something else again, a work of sensuous surfaces and vivid movement.

The prestige of bronze, fed by the demand for public monuments and a series of world's fairs and expositions that called for sculptural enhancement, remained high through World War I. It embodied all manner of concepts, from sober-sided large-scale emplacements, such as Robert Tait McKenzie's "The Youthful Franklin," 1914, a sentimental costume piece still standing at the University of Pennsylvania, to small portable sculptures for house and garden, exemplified by "Girl With Butterflies," circa 1900, a sugary figure of idealized childhood by Bessie Potter Vonnoh. Cowboys and Indians got a very big play. One of the show's liveliest works in this genre is Frederic Remington's "Trooper of the Plains," 1909, a comical tabletop study of horse and rider not far in spirit from our own Red Grooms (also in the show), and whose crisp forms and rugged textures mine the full potential of lost-wax casting. Animals alone were also a popular theme, exploited to the hilt by Anna Hyatt Huntington in "Riders to the Sea," circa 1912, a fantasy of horses breaking in a kind of Japanese wave that melds agitated forms and surfaces in a dramatically tense composition.

But by the 1920's, for artists committed to the contemporary, bronze was out. It was too closely associated with the old-fashioned Beaux Arts style, as Mr. Reynolds notes in his catalogue essay, and the notion of big, narrative public monuments had become passe. Too, artists wanted a more direct involvement with materials than the casting process provided, keeping their own hand in from beginning to end. So, while some innovators, such as Elie Nadelman — represented here by a coy, folk-artish "Female Dancer" — still included bronze in the repertory, it was largely the die-hard conservatives who continued to work it as a major medium. One was the academic Paul Manship, whose archaized mythical figures look very quaint today. His "Moods of Time: Day," 1938, one of four fountain pieces for the 1939 New York World's Fair, shows Day as a running figure with the sun held out in his hands, accompanied by a small horse on a cloud of solid bronze.

• • •

Yet Manship's one-time assistant, Gaston Lachaise, emerged from academicism to find his own powerful voice. His well-known "Peacocks," 1922, a brilliantly stylized work in which three peacocks are symmetrically united by the fan-shaped tail of one, is a masterpiece of elegant decoration. But a far cry from the classical restraint of this is "Walking Woman," another Lachaise work of the very same year, in which a figure of bright golden bronze strides vigorously ahead, the voluptuous mass of her body specifically sexual in its impact.

Bronze was briefly revived by sculptors working in the Abstract Expressionist mode right after World War II, and there are some fine early pieces by Seymour Lipton, Herbert Ferber and David Hare: strong, emotive works of biomorphic presence whose rugged surfaces intensify their Expressionist feeling. And then, with the shift to a cooler esthetic sensibility represented by Pop and Minimalism, bronze fell out of favor again, skipping a couple of generations (an important exception is shown here, a beautiful 1969-70 example of Tony Smith's stark, multifaceted geometric pieces, of low-key brown-black patination, set in the museum's back garden).

That it's currently in another revival stage — certainly due to the Expressionist nature of much current art — is left in no doubt by some of the wonderful contemporary pieces in the show. On the more traditional side are Harold Tovish's "Ceremonial Axe-Head," a small work suggesting an elegantly flattened human skull, and Howard Newman's macabre "Recumbent Head," another

Continued on Page 31

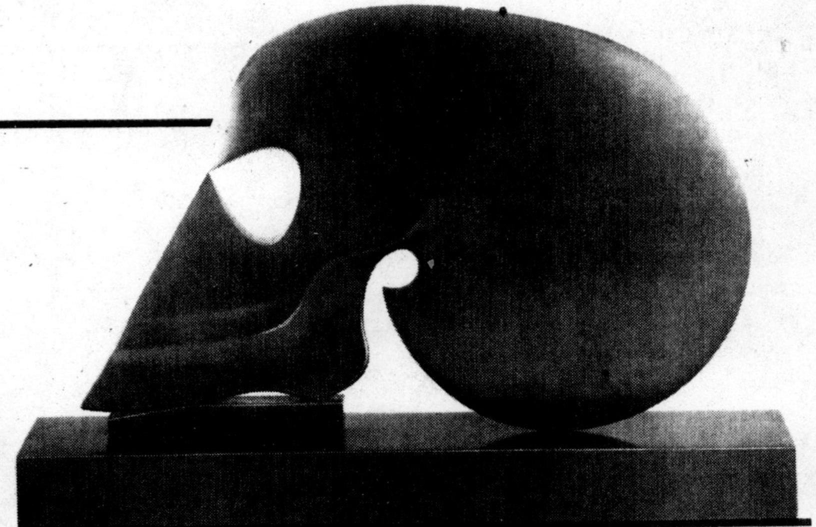

Harold Tovish's "Ceremonial Axe-Head" (1975) in the exhibition of "American Bronze Sculpture" by 57 artists at the Newark Museum

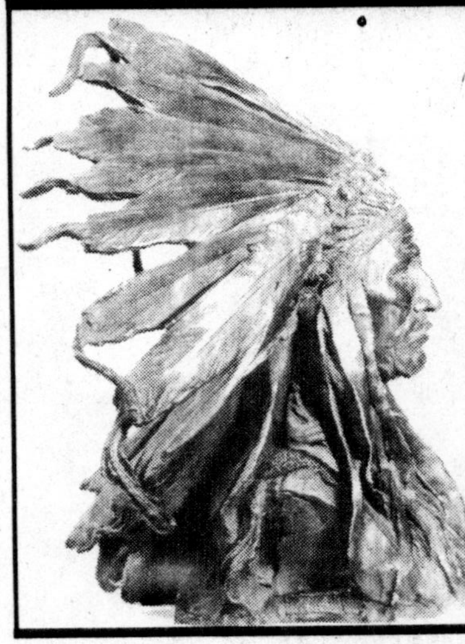

Adolph Alexander Weinman's "Chief Blackbird, the Ogalla Sioux" (1903)— American sculptors once gave cowboys and Indians a very big play.

NANCY GRAVES
Born 1940, Pittsfield, Mass.
Lives in New York City

Graves received her B.A. from Vassar College and a B.F.A. and M.F.A. from the School of Art and Architecture, Yale University, in 1964. After completing her studies, the artist received a Fulbright-Hayes Grant in painting to work in Paris. She spent the following year living in Florence. Since the late 1960s, Graves has used natural forms as the subject matter of her sculpture, painting, drawing and film. To address the formal concerns of her art she has dissected her subjects, analyzed individual parts and details, and reconstructed them in abstract, non-illusionistic, ways. The subjects of her investigations have ranged from camels to bathymetric maps of the ocean floor. Since about 1980, Graves has explored these ideas in a series of brightly painted bronze sculptures. Most often these have been created out of a great variety of plant materials (leaves, pods, stems, etc.) that are individually cast, combined with casts of other objects, welded into a single piece and, finally, painted by the artist. Like many contemporary sculptors, Graves utilizes some of the traditional qualities of bronze—its permanence and ability to reproduce intricate surfaces—but, by hiding it under a coat of paint, does not allow the material to play an important role in how we read the image.

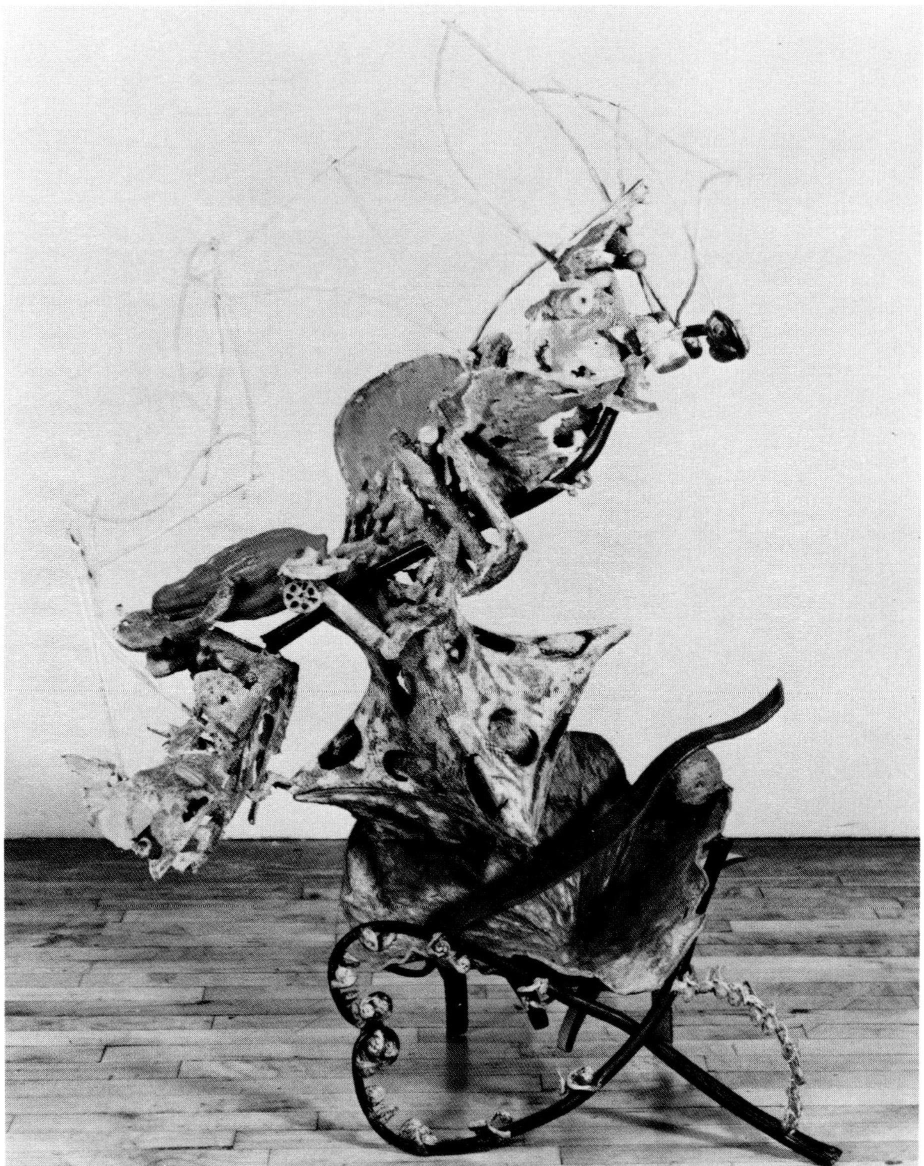

83.
Lacinate
1983-84

Cast and welded bronze, painted, 54″ x 48″ x 41″

Lent by Edna S. Beron

Lacinate is a botanical term refering to leaves with deep irregular lobes.

JOEL SHAPIRO
Born 1941, New York City
Lives in New York City

Shapiro studied at New York University, receiving a B.A. in 1964 and an M.A. in 1969. Since he began exhibiting in the late 1960s, the artist has worked primarily as a sculptor, in a variety of materials. Shapiro's work has made him one of the most provocative and significant artists of his generation and a leader of what has been termed Post-Minimalism, art created since the early 1970s that attempts to address the division between abstraction and representation. For the catalogue of his 1982 retrospective at the Whitney Museum of American Art, Shapiro wrote of his work, "I'm more interested in how one thinks about something than in what something looks like. I am interested in what a house or a figure might mean, or what it means to me. I am interested in my capacity to refer to it in terms of sculpture, but not to illustrate it or describe it."

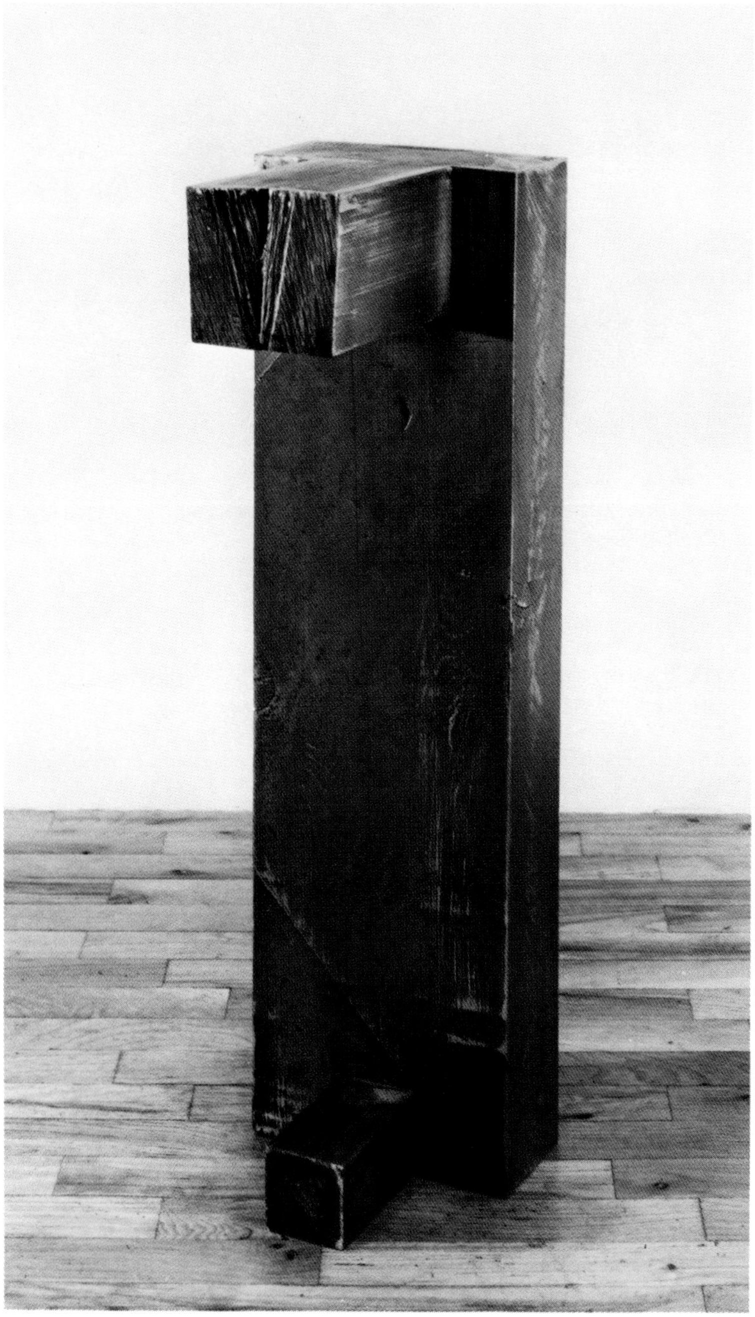

84.
Untitled
1983-84

Cast bronze, 30″ x 8⅝″ x 8½″

Signed: *Shapiro* (underneath)

Inscribed: *2/2 83-84* (stamped underneath)

Lent courtesy of Paula Cooper Gallery, New York

This piece was constructed of lumber and cast in an edition of two, of which this is the second, at the Johnson Atelier in Princeton, New Jersey. The small edition and the use of the sand casting method were of special concern to the artist.

LYNDA BENGLIS
Born 1941, Lake Charles, La.
Lives in New York City

After graduating with a B.F.A. from Newcomb College, New Orleans, in 1964, Benglis moved to New York City and attended the Art School of The Brooklyn Museum. She has worked in a variety of media, ranging from poured beeswax and colored polyurethane to metal coated wire. In 1971, the artist exhibited her series of wall sculptures constructed of cloth over wire armatures and formed into the shapes of knots. In the mid-1970s, she created sculptures of poured polyurethane foam that were cast in bronze or other metals and had a dense, heavily-layered appearance. At present, Benglis is best-known for metallic-finished wall pieces that explore various combinations of pleated, folded and knotted forms. The artist has been active in the Feminist Art movement and her work has frequently contained strong sexual-political messages.

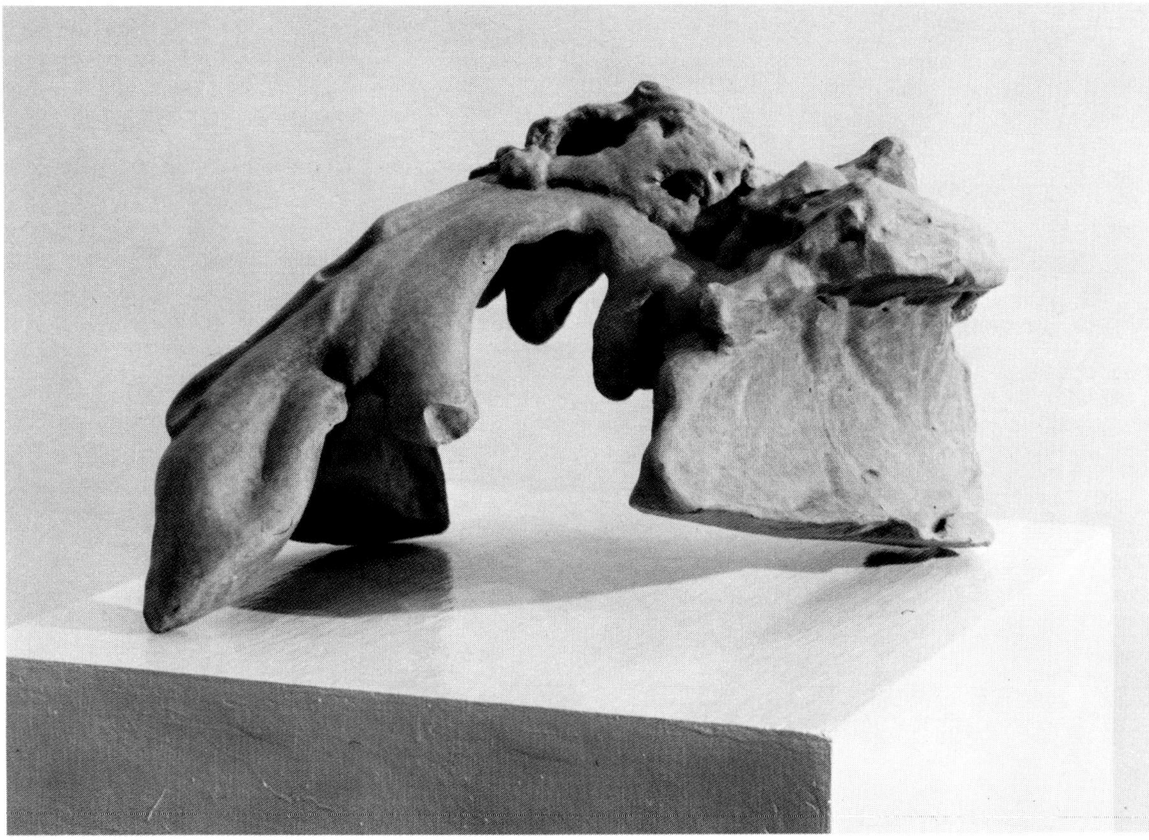

85.
Wave
1977-78

Cast bronze, green patina,
11″ x 18″ x 16″

Signed and dated: *Lynda/Benglis/ 1977-78/ 1/3* (underneath)

Lent courtesy of Paula Cooper Gallery, New York

This piece was the first casting in an edition of three. A larger, and closely related, sculpture was commissioned as a fountain for the 1984 World's Fair in New Orleans.

HERK VAN TONGEREN

Born 1943, Holland, Mich.
Lives in Hopewell, N.J.

Since 1977, Van Tongeren has directed the Johnson Atelier, an important foundry and teaching facility located in Princeton, New Jersey. Trained at the University of Colorado, Boulder, where he received an M.F.A. in 1970, he is one of the few contemporary sculptors who is a skilled craftsman in the casting process. The artist has also taught at the University of North Carolina, Greensboro (1971-74) and San Diego State University (1975-77). Van Tongeren's sculptures are usually composed of simple geometric or architectural forms arranged in a stage-like space. Although they have often been compared to the early Metaphysical paintings of Giorgio DeChirico, the sculptor tends to speak of his work more in terms of the formal relationships of objects within a controlled space. It is the tension set up between these objects and the artist's manipulation of perspective that gives his work its sense of mystery. Although best known for his bronze sculptures, Van Tongeren is also currently working in highly polished marble.

86.
Teatro IX
1982

Cast bronze, mottled green patina, 17⅛″ x 24⁷⁄₁₆″ x 12⅛″

Signed and dated: *Herk Van Tongeren 2/7 82′*/© (on back at right)

No foundry mark (cast at Johnson Atelier, Princeton, New Jersey)

Collection of The Newark Museum. Purchase 1983 The Members' Fund (83.69)

This is the second casting in a planned edition of seven.

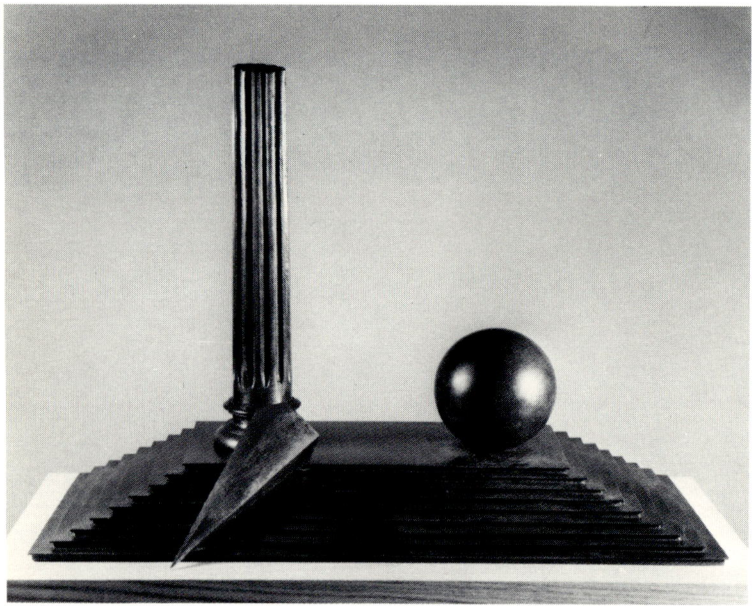

HOWARD NEWMAN

Born 1945, Elizabeth, N.J.
Lives in Newport, R.I.

Newman attended Miami University of Ohio and Rhode Island School of Design, where he received an M.F.A. in 1971. That same year he received a Fulbright Grant for study in Italy. Newman works in an idiom that is both abstract and figurative. The torsos and heads that comprise the subjects of much of his sculptures are composed of tightly interlocking geometric shapes that give them a fantastical, almost science-fiction, look.

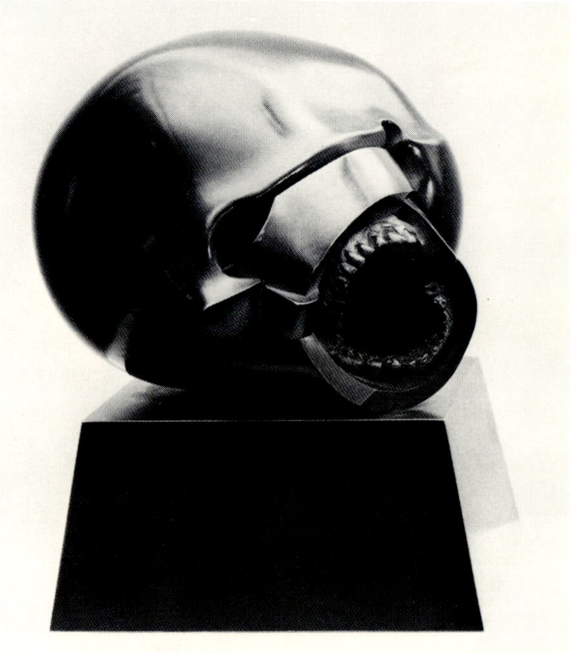

87.
Recumbent Head
1975-83

Cast bronze, dark green patina, 10″ x 8″ x 7″ (including integral base)

Signed and dated: © *H. Newman 1983. 2/9.* (on base at back right)

Lent by Cordier & Ekstrom Gallery, New York

BRYAN HUNT

Born 1947, Terre Haute, Ind.
Lives in New York City

The youngest artist in the exhibition, Hunt received his B.F.A. from Otis Art Institute in Los Angeles in 1971. Since his 1974 solo exhibition at the Clocktower (Institute for Art and Urban Resources) in New York, the artist has been included in many shows of new bronze sculpture. The attenuated forms of his recent sculptures are based on a waterfall motif that Hunt has developed in his work since 1975. These forms often have figurative as well as landscape references. However, the artist's interests are more formal than illustrative, with the gestural lines and shallow three-dimensionality of his sculptures creating "drawings in space."

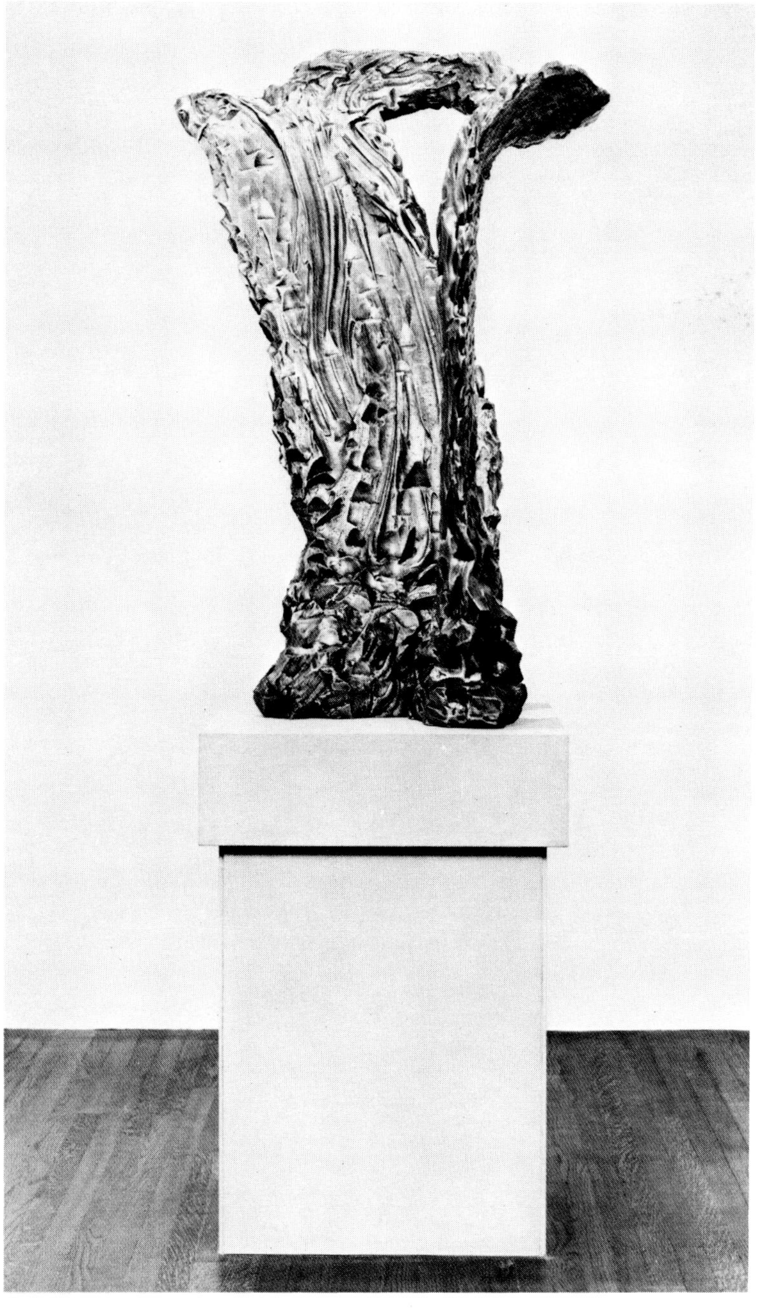

88.
The Cloak of Lorenzo
1980-81

Bronze, dark brown patina, on limestone base, 92″ x 34¼″ x 28½″ (including base)

Lent courtesy of Blum Helman Gallery, Inc., New York

Broder, Patricia Janis. *Bronzes of the American West.* New York: Harry N. Abrams, 1974.

Caffin, Charles H. *American Masters of Sculpture.* Garden City, N.Y.: Doubleday, Page & Co., 1913.

Craven, Wayne. *Sculpture in America.* New York: Thomas Y. Crowell Co., 1968.

Fusco, Peter and H.W. Janson, eds. *The Romantics to Rodin. French Nineteenth-Century Sculpture from North American Collections.* Los Angeles: Los Angeles County Museum of Art in association with George Braziller, Inc., 1980.

Gardner, Albert Ten Eyck. *American Sculpture: A Catalogue of the Collection of the Metropolitan Museum of Art.* New York: The Metropolitan Museum of Art, 1965.

Geldzahler, Henry, ed. *New York Painting and Sculpture, 1940-1970.* New York: E.P. Dutton & Co., Inc., 1969.

Gerdts, William H. *A Survey of American Sculpture: Late 18th Century to 1962.* Newark: The Newark Museum Association, 1962.

Pincus-Witten, Robert. *Postminimalism: American Art of the Decade.* New York: Out of London Press, 1977.

Proske, Beatrice Gilman. *Brookgreen Gardens.* Brookgreen Gardens, S.C.: Brookgreen Gardens, 1968.

Shapiro, Michael Edward. *Cast and Recast. The Sculpture of Frederic Remington.* Washington, D.C.: Smithsonian Institution Press, 1981.

_________. "The Development of American Bronze Foundries, 1850-1900." Ph.D. Dissertation, Harvard University, 1980.

Stover, Donald L. *American Sculpture. The Collection of The Fine Arts Museums of San Francisco.* San Francisco: The Fine Arts Museums of San Francisco, 1982.

Taft, Loredo. *The History of American Sculpture.* 1903. Reprint. New York: Arno Press, 1969.

200 Years of American Sculpture. New York: David R. Godine, Publisher, in association with the Whitney Museum of American Art, 1976.

Vanguard American Sculpture 1913-1939. New Brunswick, N.J.: Rutgers University Art Gallery, 1979.

Wasserman, Jeanne L., ed. *Metamorphoses in Nineteenth-Century Sculpture.* Cambridge, Mass.: Fogg Art Museum, distributed by Harvard University Press, 1975.

Note:

Currently there is no single book that covers the history of American bronze sculpture. The works listed in this general bibliography discuss the major artists and topics included in the exhibition. The reader is also referred to the many existing monographs on individual artists. Of particular value to the author were the published writings of Michael Shapiro.

Adams, Herbert, p. 20
Ball, Thomas, p. 14
Barthé, Richmond, p. 53
Baskin, Leonard, p. 59
Beach, Chester, p. 41
Benglis, Lynda, p. 67
Bitter, Karl, p. 29
Borglum, Gutzon, pp. 30-31
Borglum, Solon, p. 31
Brown, Henry Kirke, p. 13
Burroughs, Edith Woodman, p. 33
Dallin, Cyrus Edwin, p. 23
Davidson, Jo, p. 46
Deming, Edwin Willard, p. 21
Diederich, William Hunt, p. 46
Donoghue, John Talbott, p. 19
Eberle, Abastenia St. Leger, pp. 38-39
Ferber, Herbert, pp. 54-55
Ferguson, Duncan, p. 53
Flanagan, John, p. 27
French, Daniel Chester, pp. 18-19
Friedlander, Leo, p. 49
Frishmuth, Harriet Whitney, p. 40
Graham, Robert, p. 63
Graves, Nancy, p. 65
Grippe, Peter J., p. 56
Grooms, Red, p. 62
Hare, David, p. 58
Haseltine, Herbert, p. 36
Hoffman, Malvina Cornell, p. 48
Hosmer, Harriet, p. 15
Howard, Cecil, p. 42
Hunt, Bryan, p. 69
Huntington, Anna Vaughn Hyatt, pp. 34-35
Hyatt, Anna; see Huntington, Anna Vaughn Hyatt
Jennewein, Carl Paul, pp. 50-51
Kalish, Max, p. 52
King, William, p. 61
Lachaise, Gaston, pp. 43-44
Lipchitz, Jacques, p. 51
Lipton, Seymour, p. 54
MacMonnies, Frederick William, pp. 25-27
MacNeil, Hermon Atkins, p. 28
McKenzie, Robert Tait, p. 29
Manship, Paul Howard, p. 47
Nadelman, Elie, p. 45
Nakian, Reuben, p. 52
Newman, Howard, p. 68
Potter, Bessie; see Vonnoh, Bessie Potter
Proctor, Alexander Phimister, p. 24
Remington, Frederic Sackrider, p. 22
Rhind, John Massey, pp. 20-21
Rumsey, Charles Cary, p. 39
Safferson, Matthew, p. 56
Saint-Gaudens, Augustus, pp. 16-17
Schreyvogel, Charles, p. 23
Segal, George, p. 60
Shapiro, Joel, p. 66
Smith, Tony, p. 57
Strider, Marjorie, p. 64
Tovish, Harold, p. 58
Van Tongeren, Herk, p. 68
Vonnoh, Bessie Potter, p. 34
Ward, John Quincy Adams, p. 15
Warner, Olin Levi, p. 16
Weinman, Adolph Alexander, p. 32
Whitney, Gertrude Vanderbilt, p. 37
Witkin, Isaac, p. 61
Woodman, Edith; see Burroughs, Edith Woodman
Wright, Alice Morgan, p. 41
Young, Mahonri Mackintosh, p. 38
Zorach, William, p. 50

PHOTOGRAPHY CREDITS

Cover photograph & catalogue no. 18, William G. Wagner.

Catalogue nos. 4, 28, 29 & 45, David Allison, Photography.

Catalogue nos. 5, 11, 12, 16, 20, 32, 33, 36-38, 43, 47-49, 51, 54, 61, 63, 70, 72, 74, 76, 79, 80 & 86, Armen Shamlian, Armen Photographers.

Catalogue nos. 13 & 57, Helga Photo Studio.

Catalogue nos. 23 & 41, Scott Hyde.

Catalogue no. 26, Sotheby Parke Bernet.

Catalogue nos. 52, 62, 84 & 85, Geoffrey Clements.

Catalogue no. 64, IXL Photo Co.

Catalogue no. 68, Rudolph Burkhardt.

Catalogue no. 77, Eric Pollitzer.

Catalogue no. 81, Bevan Davies.

Catalogue no. 12, Erik Landsberg.

Catalogue no. 87, eeva-inkeri.

Catalogue no. 88, Jon Abbott.

Colophon

This publication was coordinated at The Newark Museum by Mary Sue Sweeney, Director of Publications. It is designed by Johnson & Simpson Graphic Designers, Newark, New Jersey. The type is set in Palatino and ITC Franklin Gothic Demi by Elizabeth Typesetting Company, Kenilworth, New Jersey. The book was printed by Harvard Printing Company, Orange, New Jersey.